# CLIMATE OF HOPE

# CLIMATE

## OF

# HOPE

## HOW CITIES, BUSINESSES, AND CITIZENS
## CAN SAVE THE PLANET

## MICHAEL BLOOMBERG
## CARL POPE

ST. MARTIN'S PRESS ≋ NEW YORK

www.stmartins.com

Designed by Steven Seighman

The Library of Congress Cataloging-in-Publication Data is available upon request.

ISBN 978-1-250-14207-8 (hardcover)
ISBN 978-1-250-14209-2 (e-book)

Our books may be purchased in bulk for promotional, educational, or business use. Please contact your local bookseller or the Macmillan Corporate and Premium Sales Department at 1-800-221-7945, extension 5442, or by e-mail at MacmillanSpecialMarkets@macmillan.com.

First Edition: April 2017

10  9  8  7  6  5  4  3  2  1

# CONTENTS

# PREFACE

It's easy to be despondent about climate change. The issue received almost no attention during an eighteen-month presidential campaign that tested everyone's patience. As a candidate, Donald Trump said he would "cancel" the 2015 Paris Climate Agreement and roll back many of the Obama administration's efforts to reduce greenhouse gases. His pick to lead the Environmental Protection Agency is known primarily for fighting efforts to tackle climate change. Abroad, Brexit has created questions about Europe's ability to unite against this formidable problem. Yes, the Paris Agreement was a major breakthrough in climate diplomacy, but there is nothing enforceable about it—no real penalties for countries that don't honor their commitments. On top of all that, the difficulty of preventing the Earth's temperature from rising sounds enormous enough to make many people give up and hope for the best.

We see it differently. Through our work with cities, businesses, and communities, we believe that we are now in a better position to stop climate change than ever before. And that in the years ahead, forces outside of Washington can and must deliver new levels of progress.

Let's start with some perspective. It has been more than a decade since a hit film, *An Inconvenient Truth*, exposed a mass audience to the dangers of climate change and warned of a dark and frightening future. "Humanity is sitting on a time bomb," an advertisement for the film read, continuing:

> *If the vast majority of the world's scientists are right, we have just ten years to avert a major catastrophe that could send our entire planet into a tailspin of epic destruction involving extreme weather, floods, droughts, epidemics and killer heat waves beyond anything we have ever experienced.*

Well, the ten years have come and gone, but the window for action has not closed, and all is not lost. Far from it. In fact, over the past decade we have made critically important progress that allows us to see something that we couldn't in 2006: a realistic path to victory.

The two of us are deeply optimistic about the world's ability to travel that path, but for reasons that rarely generate much public attention. In the United States, most media coverage of climate change focuses on Washington, where the debate has barely budged over the past twenty years. Democrats are still accusing Republicans and their allies in the fossil fuel industry of ignoring a catastrophe in the making, while Republicans accuse Democrats and their allies in the environmental movement of scaremongering and overstating the science. There is some truth on both sides, but it is obscuring a larger truth: Cities, businesses, and citizens are increasingly taking action on their own, and they have only just begun to fight.

The two of us have no interest in the tired old debate that is still playing out on Capitol Hill. Our interest is not in winning an argument or an election. It's in saving lives, promoting prosperity, and stopping global warming.

We are writing this book because we believe that it's time for a new type of conversation about climate change that reverses all the usual ways of looking at the issue. Instead of debating long-term consequences, let's talk about immediate threats. Instead of arguing about making sacrifices, let's talk about how we can make money. Instead of pitting the environment versus the economy, let's consider market principles and economic growth. Instead of focusing on polar bears, let's focus on asthmatic children. And instead of putting all hope in the federal government, let's empower cities, regions, businesses, and citizens to accelerate the progress they are already making on their own.

We believe that by changing the way we think and talk about climate change, we can lower the temperature of the debate—and accomplish a whole lot more. Cooler heads can produce a cooler world.

Creating a new type of climate conversation begins with reconsidering the problem itself. Political and environmental leaders both tend to talk about climate change as a single massive problem, and one that only global treaties can solve. But consider this: When scientists set out to rid the planet of disease, they don't attempt to cure every disease at once, and they don't expect a single research team to come up with all the answers. Instead, by working in many different labs all over the world and sharing their discoveries, teams of scientists zero in on a single type of disorder, study its characteristics, research its causes, and experiment with cures. Meanwhile, other teams of scientists tackle other diseases. It's a strategy that has led to countless breakthroughs—and it's the same approach that we should be taking with climate change.

The changing climate should be seen as a series of discrete, manageable problems that can be attacked from all angles simultaneously. Each problem has a solution. And better still, each solution can make our society healthier and our economy stronger.

That's worth repeating: *Each part of the problem of climate change has a solution that can make our society healthier and stronger.* That idea—which threads through this book—has been largely missing from the political debate over climate change. In the United States, the debate has too often centered on how likely various doom-and-gloom scenarios are, and when they'll occur. But scaring people doesn't work.

A 2009 study concluded that fear does not, in fact, galvanize people to fight climate change. "Although shocking, catastrophic, and large-scale representations of the impacts of climate change may well act as an initial hook for people's attention and concern, they clearly do not motivate a sense of personal engagement with the issue and indeed may act to trigger barriers to engagement." Other studies have found that using exclusively dire messages about climate change can actually increase skepticism and denial of the problem.

Advocates for climate action also have a tendency to focus on abstract ideas that hold little meaning for most people, such as whether we need to limit eventual warming to 2 degrees Celsius above pre-industrial levels by 2100, or whether the upper limit should be 1.5 degrees. The reality is that the uncertainty involved means we do not know with precision whether we can achieve either of those goals; nor do we know for certain what the consequences will be if we do not. Regardless, most people put little faith in projections that far out, which is understandable. Scientists have been wrong plenty of times before. What people want to know is not exactly what will happen to the Earth eighty years from now but what will happen to their house, their job, and their community this year.

Telling people that they might possibly save the Earth from distant and uncertain harm is not a great way to convince them to support a particular policy. But what happens when you tell people that they can definitely, right now, reduce the number of asthma attacks suffered by children, save their own families and friends from respiratory disease, extend their own life expectancy, cut their own energy bills, make it easier for them to get around town, improve their quality of life, increase the number of jobs in their community, and strengthen our national energy security—all while increasing the long-term stability of the global climate?

Now that's a different story. And it's our story, too.

The benefits of taking action on climate change are tangible, immediate, and vast—but too few political leaders are talking about them. In our experience, fighting climate change goes hand in hand with improving public health, strengthening economic growth, and raising living standards—and when leaders focus on those benefits, the public responds favorably. People don't need a lecture on what may happen to the Earth one hundred years from now, when we'll all be dead and gone. They need good reasons to support change today.

The two of us have arrived at this conclusion from very different places—and from very different backgrounds. After college, one of us entered the Peace Corps; the other went to Harvard Business School. One of us spent decades in California working at and then leading the Sierra Club; the other spent fifteen years on Wall Street before founding and running a company and later serving for three terms as mayor of America's largest city. In 2004, one of us coauthored a book attacking George W. Bush for his environmental policies; the other voted for him.

Yet we approach the challenge of climate change from the same general perspective. We don't agree on every point—what New Yorker and Californian do?—but we share a strong sense of responsibility for taking on this fight, and a deep sense of optimism that we can win it, even in the current political climate. This book is our effort to convince more people—of equally diverse backgrounds, and of all political persuasions—to join us.

# PART I

# COMING TO CLIMATE

I

# GOING AFTER GOLIATH

*In preparing for battle I have always found that plans are useless, but planning is indispensable.*
—Dwight D. Eisenhower

November 2, 2004. The exit polls are exhilarating. They show John Kerry ousting George W. Bush from the White House—a goal that has been the top political objective of tens of thousands of Sierra Club volunteers, leaders, staff, and donors since Bush opened his war on the environment in the spring of 2001. As the executive director of the club, I've seen our team pour their hearts and souls into a groundbreaking new progressive coalition determined to reach and motivate anti-Bush voters who might otherwise not get to the polls, and it seems to have worked! One of the major funders of the effort, George Soros, is hosting an election evening party in Manhattan, and the room is buzzing.

For a while. As real votes replace exit polls, things get tighter and tighter. The race finally comes down to Ohio, and late in the evening it is clear that even though thousands of voters all over Ohio stood in the rain to vote for Kerry, thousands more came out for Bush, and Kerry has lost Ohio, and the White House, by 2.1 percent. After thirty-five years as a practicing environmental advocate, ten of them as the Sierra Club's executive director, I am at a loss. I came of age in the environmental movement at a time when its concerns were shared by most Americans of both parties. A bipartisan Congress and White House created the Environmental Protection Agency, passed the Clean Air and Clean Water Acts, and expanded parks and wilderness. In the decades that followed, however, this bipartisanship faded, and the Sierra Club faced an increasingly hostile Republican party. We weathered Ronald Reagan's interior secretary, James Watt, and his contempt for the Grand Canyon, which he called "boring." We helped the country withstand Newt Gingrich and his campaign to shred the public health safety net. But nothing compared to Vice President Dick Cheney's brand of predatory disregard for clean air, water, and landscapes, which had also come to dominate the Republican Party. And now the party had just recaptured full control of the White House and both houses of Congress. How should the Sierra Club's leadership respond? What is our strategy for coping with a second Bush administration?

Defense, it is clear, is not enough. To find an answer, we decided to mount the most intensive consultation process in the club's modern history, inviting over five thousand grassroots leaders to participate in a series of meetings, surveys, and discussions. It culminated in our first national convention, which we held in San Francisco in September 2005.

Even before the convention, the results from the state and local consultations suggested a surprising shift in the membership's concerns— after more than a hundred years in which the protection of wild places was our highest priority, club leaders were now saying that climate change needed to be at the top of our agenda. Then, even as the delegates were flying in to San Francisco, I got a call from Al Gore. In fact, we had invited Gore to speak at the convention, but he was already booked to address the National Association of Insurance Regulators at their meeting in New Orleans. Now he was calling to say that, due to a hurricane headed toward the Gulf Coast, the meeting had been canceled. Would we still like him to appear?

The answer was yes, of course. And so, as Hurricane Katrina bore down on New Orleans, the former vice president shared with the club's leadership the slide show that became the basis for *An Inconvenient Truth*.

The impact of Katrina and Gore's stunning presentation only strengthened our commitment to make climate our priority. After the convention, as executive director, I had a new mission.

Admittedly, climate change was not entirely new territory for us. The club had worked for years on energy and climate, taking the lead in pressuring the auto industry to improve fuel economy and cut carbon dioxide emissions from cars. But it had never been our top priority. How does a grassroots activist organization in the United States set out to stop global warming?

I didn't know it yet, but this new priority would mean enormous changes in the way I saw the world and environmentalism. Previously environmentalists worked to stop bad things—pollution, clear-cutting, overfishing—but we more or less accepted the big-picture American economy, with the established industries that made it up. Not anymore. Now we were about to find ourselves in a different business: helping to foster a different kind of economic development, one based on knowledge and technology rather than fossil fuels. After thirty-five years of working to clean up after twentieth-century industrialism, environmentalists were about to plunge into creating its twenty-first-century replacement. But before we could go full tilt toward the new, we had to stop the last spasms of the old—an energy future crafted during George W. Bush's first term by Vice President Cheney.

Staff and volunteer leaders began brainstorming, and eighteen months later, in February 2007, about one hundred club leaders gathered in Tucson in an unconventional forum to decide which campaigns we would rally around.

Anyone in the group who had a campaign idea he or she wanted to pitch set up a whiteboard in a corner. The rest of the group then "voted with their feet," testing out different conversations until they found the one that gripped their imagination. At the end of the session we had four or five lively groups, but one had managed to attract almost half of the total audience. A club lawyer from the Midwest, Bruce Nilles, was proposing that the Sierra Club target the linchpin of the Bush administration's energy proposal: building more than 150 new coal-fired power plants. Nilles pointed out that coal was the biggest source of climate pollution in the United States. Not only that, the proposed new plants would emit so much climate pollution for the forty years of their expected lifetime that if they were built, it would become mathematically impossible to tame the global-warming monster. This next generation of coal plants, he told us, would lock the United States into 750 million additional tons of carbon dioxide emissions every year, just when we needed to be cutting those emissions by that same amount by 2012.

Nilles' pitch was that we faced what I have heard the Orthodox Patriarch Bartholomew call a *kairos*: a supreme moment at which one simply must act, however implausible or inconvenient. His impassioned plea struck a deep chord with the club's leaders, but the ambition of the task left me uneasy. Afterward, I took him aside. How could we take this on? What was his strategy?

"We just fight every single new coal plant," he replied.

"Who is going to help us?" I asked.

"I have no idea if anyone will," he admitted. "Most groups working on this just want to challenge one or two to try to get them a little cleaner."

I was not reassured, but the die was cast.

The Sierra Club was now committed to transforming the entire energy sector of the United States, and to prevent it from locking itself into another generation of coal-fired power.

Thanks to some initial gifts from small family foundations and the leaders of the nascent solar industry, Bruce was able to hire lawyers for the effort, which we called "Beyond Coal," and they began looking for planned coal-fired power plants to challenge. What they uncovered was shocking. The relationship between the coal industry and the electric utilities was so incestuous that coal executives simply assumed their plan would be rubber-stamped. Even more disturbing, the government regulators who were supposed to protect consumers and the environment were in many instances virtual partners in the planned coal rollout. None of these parties had the slightest experience of being challenged by citizens, and certainly none of them expected the next generation of coal plants to face serious opposition when they came before regulators for approval.

In fact, the coal team found two plants in the Midwest that were well under construction, with more than $100 million in expenses incurred—even though the appropriate government agencies had never issued the required permits for construction. When the club's lawyers inquired of one utility why it had assumed that it could build a huge power plant without permits, the response was, "We asked the regional office of the EPA and they told us not to bother with permits."

Our lawyer's response: "Let's see what a federal judge thinks about that theory." Unsurprisingly, the judge told the utility lawyers that he was perfectly willing to hear the case, but he strongly advised them to settle it instead.

The companies building the two projects were now desperate. They didn't want to admit to their shareholders and customers that each had squandered more than $100 million on projects that were now almost certain to be rejected by a court. They asked us if we really wanted to see that much money go to waste. Bruce and our coal team, mindful not to overreach, offered a deal: If the companies would shut down their oldest, very dirty coal plants and purchase an equal volume of clean wind power to lower their carbon pollution, the club would let them finish the ones that were half built. The companies agreed to the deal. It was a win for them and for the environment, and particularly for the people forced to breathe the filth from the old plants. By the spring of 2007, only a few months after the inception of Beyond Coal, the club had scored two stunning victories.

News that *two* companies were shutting down old plants, buying thousands of megawatts to jump-start the Midwestern wind power industry, and signing an agreement with the Sierra Club sent shock waves throughout the utility industry and its financiers on Wall Street.

A lobbyist for the biggest new coal plant in the queue, the 2,100-megawatt Sunflower plant in Kansas, approached me at a Washington cocktail party. She knew me from her time in the Clinton White House. "We'd like to explore having you green-light Sunflower if we shut down some of our old coal," she explained. "Sorry," I replied. "That was a Washington's Birthday special only." Sunflower was not even close to being built—there was no reason to give it a pass. With the club leading the opposition, the Kansas Department of Health and Environment eventually rejected its permit. This led coal proponents to launch a media campaign against Governor Kathleen Sebelius, linking her to Vladimir Putin, Hugo Chávez, and Mahmoud Ahmadinejad of Iran. The *Washington Post* disparaged the ads as "extremely misleading." Sebelius's successor tried to revive the plant but failed.

These three early victories were encouraging, but three is not 150. The club's resources, and Bruce's team, were far too small to scale up the challenge and accomplish our goal. The large foundations interested in the coal fight rejected our goal as implausibly ambitious; instead they focused their efforts on a few plants only, hoping to settle for better pollution controls. The idea that we were on the cusp of a new era in which cleaner fuels would replace coal completely was too foreign for most of the players to grasp.

Later in the spring of 2007 I got a phone call. Aubrey McClendon, the head of a natural gas company called Chesapeake Energy, wanted to meet with us. He was frustrated that newly expanding supplies of natural gas, which would generate far less air pollution than coal, were being kept out of the market by the coal plant boom. He had already intervened in Texas and Oklahoma against two coal plants. After noting the club's presence in a number of other battles, he wanted to make a donation.

It turned out that he wanted to make a big donation—$5 million the first year. Aubrey wasn't eager to have his coal competitors on his back, so he made the gift anonymously. The coal industry suddenly had a much bigger fight on its hands.

Six months later, I attended my first United Nations Climate Change Conference, in Bali, Indonesia. The political director of the International Brotherhood of Boilermakers, Abe Breehey, who enjoyed a wonderful reputation among my labor friends, was also there, and he asked me to have a drink with him. Abe had heard about our fight against coal and he congratulated me on our project. "I need to explain one thing," he told me. "We build boilers. We don't care what powers them. So we're not pro-coal. We are pro-boiler." But he wondered how far we could take this fight. "There are 150 in the queue," Abe continued. "You can't possibly go after them all, whatever your press releases say."

I paused. "Actually, as of a few months ago, we now have the resources and the staff to challenge every single one—and we plan to do so."

He put his glass down. "Well," he said, "it looks like we and the utilities need a new business model." If the coal plants weren't going to get built, boiler makers needed to make sure that natural gas plants would be. In other words, Abe immediately grasped what the utilities refused to accept: Coal's heyday was over.

For the next three years utilities struggled to get regulators to allow them to build scores of new plants. Ultimately, of the 150 coal plants that had been queued up when Bruce Nilles stood before his whiteboard in Tucson, only 30 were ever built. We stopped 100,000 megawatts of new coal power. To put it another way: Had those plants been built, they would have increased America's coal power production by 30 percent—and locked in at least another generation of pollution and carbon emissions. Of the 30 plants that did get built, almost all of them turned into economic white elephants, driving up utility rates, bankrupting companies and communities, and in some cases sitting idle because no one could afford to operate them.

Coal power was an idea whose time had come and gone. People just didn't know it yet.

While climate change had become the club's top priority, and Beyond Coal the most successful anti-pollution campaign we had ever run, we were working on a broad range of other issues as well. In the spring of 2007, I received a second fortuitous phone call, this one from New York City Mayor Michael Bloomberg's office. One of his deputy mayors, Kevin Sheekey, reached out to see if we could help with PlaNYC, the city's new sustainability plan. In particular, he was seeking support for a proposal to use "congestion pricing"—a toll charged during peak driving hours—as a mechanism to raise funds for improving mass transit. Better transit is a key climate strategy as well, so we were eager to help. My first assignment was to help persuade New York Governor Eliot Spitzer to support congestion pricing. (Eventually he did, although I doubt my lobbying was the key.) More significant, however, was that a partnership between the club and the mayor took root, while proposed coal plant after proposed coal plant bit the dust.

Those of us in the Beyond Coal campaign began to realize that we were on the way to something big—completely ending the U.S. coal boom. The combined impact of the club's challenges, Wall Street's nervousness about whether new coal plants would actually get built, and the declining price of natural gas brought on by the shale boom prompted utilities to lose interest in pursuing the black rock as their energy source.

Still, there was major work to be done. The country's existing coal plants—many of which dated from the First World War—were still the largest source of carbon pollution in the country, belching some 2 billion tons of carbon dioxide into the atmosphere each year. And so Bruce

Nilles designed a new and even more audacious game plan: We would mobilize neighbors and citizens to shut down the 535 coal boilers built before 2000—which were then providing about half the nation's electricity—and replace them with wind and solar.

In 2010, at sixty-five, I stepped down as the Sierra Club's executive director, but remained on as chairman through 2012. I had stayed in touch with Kevin, and over lunch one day, I told him about the club's new vision—Beyond Coal Phase II: Shut down the old coal plants.

Kevin shared my excitement, and soon, so would the mayor. I didn't know it then, but that lunch would launch the best-planned and most ambitious environmental campaign in the Sierra Club's—and perhaps the U.S. environmental movement's—history. This campaign yielded, as Mike will later explain, the biggest and fastest impact on the planet. It also led to our partnership, which has produced many collaborative efforts to drive progress on climate change—including, now, this book.

2

# PLANYC

*If you want to get things done, ask a mayor.*
—DENIS CODERRE, MAYOR OF MONTRÉAL, CANADA

I'm not exactly your stereotypical environmentalist. I don't own a pair of Birkenstocks, eat granola, hug trees, lie down in front of bulldozers, oppose GMOs, or lose sleep over spotted owls. I don't want to ban fracking (just do it safely) or stop the Keystone pipeline (the oil is coming here one way or another), and I support nuclear power. I've spent most of my career in finance, and the technology my company makes is used by traders, financiers, and executives around the world (the smart ones, at least). So why did a guy like me become a crusader against climate change? Very simply: to save and improve lives.

When I was elected mayor of New York City in 2001, no one was particularly focused on the environment. Terrorists had destroyed the World Trade Center two months earlier, the city was in mourning, and many people were predicting that businesses and families would flee and high crime rates would return. Getting the city back on its feet—emotionally and economically—and proving the naysayers wrong was my number one priority.

I understood why people feared that New York might slip back into disorder and decay. I had moved to New York City in 1966, after graduating from Harvard Business School, just as the city was beginning its descent into a spiral of crime, drugs, abandoned buildings, crumbling infrastructure, filthy streets, graffiti-covered subways, lost manufacturing jobs, worsening racial tensions, and terrible air pollution. The mayor at the time, John Lindsay, used to joke, "I never trust air I can't see." The middle-class dream was to escape the mess and move to the suburbs. Many did.

During the 1970s, New York City lost more than 10 percent of its population, and it was far from alone. Cities across the country were hemorrhaging jobs and people. At the same time, urban factories were closing and leaving a legacy of environmental degradation, with poisoned soil and water. The question back then wasn't whether cities could be saved, but whether they were even worth saving. And many thought that they were not.

By the turn of the twenty-first century, though, a new urban renaissance was taking hold. New York and other cities were becoming magnets for young people who were attracted to the very same urban values that had drawn people for generations: culture, community, cuisine, and career opportunities, all accessible by foot or mass transit. More than ever, cities proved to be centers of innovation, diversity, and discovery. Which is why, for the first time in history, the majority of people now live in cities. By midcentury, three-fourths of the world's population are expected to be city dwellers. People are voting with their feet, and cities are winning in a landslide.

There are many benefits of this change, but I hadn't realized a critically important one before becoming mayor: Cities are actually the key to saving the planet.

One reason this urgent fact doesn't get the attention it deserves is that cities seem so contrary to nature. And yet one of the best things an individual can do to fight climate change is live in a dense urban environment. Why? Because most urban residents live in apartments that are smaller than the average American home and require far less energy to heat in winter and cool in summer. City residents also tend to drive less, because they can walk, bike, or take mass transit to get to work and get around. As a result, the average per capita carbon footprint in New York City is two-thirds smaller than the national average.

But density is not the only reason that cities are well positioned to fight climate change. Three others are just as important:

*First, as Atlanta Mayor Kasim Reed says, "Cities are on the front lines of climate change—we are where the action is."* That's true in terms of both emissions and dangers. Most climate change problems originate in cities—they account for about 70 percent of greenhouse gas emissions. By sheer virtue of numbers, city inhabitants are heating up the planet more than their rural counterparts; they are also paying a higher price for the consequences. The power plant that belches toxins into the air may be outside the city, but the people in the city are using the energy that it generates. So the amount of energy we use in cities determines how much the plant belches. Cities (as I'll describe in greater detail later) also face the worst risks from those emissions, because most of them have been built in coastal areas, where they are vulnerable to rising seas and stronger storms. As the primary drivers of climate change, cities must take the lead in tackling it. And as the likeliest victims of climate change, they have the greatest incentive to do so.

*Second, mayors everywhere tend to be more pragmatic and less ideological than national legislators.* The reason is: mayors are most directly responsible for people's well-being. They have to solve problems and deliver essential services. When children suffer from asthma attacks because of dirty air, people call on the mayor to do something about it—not their congresswoman or senator. If a mayor improves the life of a community, people don't much care what party he or she belongs to. As former New York City Mayor Fiorello La Guardia once said, "There's no Democratic or Republican way of cleaning the streets."

*Third, mayors see fighting climate change as a spur to faster economic growth.* Traditionally, national governments have seen fighting climate change as a cost that is paid for in lower economic growth; some national legislators are in favor of paying it, and some are against. Mayors, on the other hand, view the problem through a different lens, and this difference in perspective stems from a difference in responsibilities. Members of Congress spend much of their time fighting about how to redistribute money sent to Washington from cities and states. Mayors—who, unlike members of Congress, actually balance budgets—think about delivering services and improving residents' quality of life. As they do, it's natural for them to ask: What investments can we make that will allow us to attract more people and business? What services can we provide that will give us a competitive edge in the marketplace for entrepreneurs, young college graduates, and families?

In part, mayors take climate change more seriously because our model for strengthening cities has evolved. It used to be that urban economic development focused on retaining industries and luring new businesses with incentive packages. But in the new century, as businesses have become more mobile and the world more connected, a different and far more effective approach has emerged: focusing first and foremost on creating the conditions that attract people. This creates competition among cities—which of them can offer the best schools, the safest streets, the biggest parks, the most extensive mass transit, and the cleanest air? If you live in a city where your asthmatic child keeps ending up in the hospital and the city next door has cleaner air, you might call your mayor and say, "I'm not going to vote for you again unless you do something to address this."

Cities everywhere are increasingly demonstrating a phenomenon I often point out about New York: Talent attracts capital more effectively than capital attracts talent. People want to live in communities that offer healthy and family-friendly lifestyles. And where people want to live, businesses want to invest.

I wasn't surprised when, in 2015, Beijing announced that it would close its last four major coal-fired power plants: Dirty air is a major liability for a city's business environment. Any marginal financial benefit that the city's plants offered had been eclipsed by their net costs, including damaging the health of the capital's inhabitants and driving away foreign investment. Dirty air made it more difficult for Beijing to attract skilled workers and the businesses that rely on them.

Beijing is just the latest city to reduce its air pollution for economic and health reasons. *One of the biggest changes in urban governance in this century has been mayors' recognition that promoting private investment requires protecting public health—and protecting public health requires fighting climate change.* Why? Because the largest sources of air pollution that threaten public health are also the biggest sources of the greenhouse gases that are warming our planet.

The former president of the World Medical Association, Dana Hanson, called climate change a "massive threat to global health that will likely eclipse the major known pandemics as the leading cause of death and disease in the twenty-first century." He argued that "the health of the world population must be elevated in this discussion from an afterthought to a central theme." He couldn't be more correct—and more and more mayors of all political parties are making it a central focus of their work.

When I was in city hall, connecting the dots between climate and health and the economy didn't happen all at once. Instead, it was the result of a careful study prompted by a single number: one million. In 2005, the demographers in our city planning department projected that New York City would have one million more inhabitants in 2030 than it had had in 2000. That would be the equivalent of every resident of Atlanta *and* Miami—and then some—moving to New York City. Our population was already at a record high, with overcrowding a concern and much of our infrastructure a century old or more. How would we deal with another million people?

Don't get me wrong. This was a problem of success—the kind of problem every mayor wants to have. Cities are never static; they are either growing or dying. And New York had rebounded faster and stronger from the 9/11 attacks than anyone had expected. We were creating jobs and attracting businesses and people. The projected additional million was good news for the planet, too, because otherwise many of those one million people would be living in larger homes and driving more frequently. But success creates a whole new set of challenges. It was easy to envision the worst possible scenarios: Paralyzing traffic jams. Overcrowded buses and trains. Long lines for park facilities. The electrical grid overloaded. Sewage and water systems strained.

And looming over all of these challenges was the growing specter of global warming.

It's easy to forget when you're standing in the middle of Times Square, but New York is a coastal city, with 520 miles of waterfront. Only one of the five boroughs, the Bronx, isn't an island or part of an island, and even it is surrounded on three sides by water. The whole city was built around the harbor. This meant that, in addition to facing the prospect of absorbing one million more people, we would have to do it while also preparing for the possibility of more powerful storms and more destructive flooding, exacerbated by melting polar ice caps and rising sea levels. So as Congress was endlessly debating the science of global warming, while at the same time ignoring the infrastructure needs of cities, we began reimagining New York City for the twenty-first century.


Cities and nations thrive when leaders anticipate the future—and dream big. That lesson is clear from New York's history. When city leaders laid out Manhattan's street grid in 1811, nearly everyone lived in Lower Manhattan. But they stretched the grid far into what was then the countryside, because they looked decades ahead and envisioned a great metropolis. The same is true of New York's subway system, which reached northern

Manhattan at a time when it was still farmland. Our water systems, bridges, tunnels, and even our parks were built with future generations in mind. When an enormous Central Park was first proposed in the 1850s, at a time when most of Manhattan was still forests and fields, many critics argued that it was too big and too expensive. The city council even tried to reduce its size. Today it's the most famous park in the world, and there is no one who wishes the doubters had won.

After learning from our demographers about our projected growth, Dan Doctoroff, one of my deputy mayors, convened a working group to figure out where we might house one million more people. The group intended to create a long-term land use plan, but as they began their work, they realized the complexity of the task at hand. They couldn't plan for new housing without also thinking through the needs that new housing would create and the new challenges it might present. Where would children go to school? What parks would families visit? How would they get around the city? Where would we put all the waste that they would create? We couldn't think about any of these issues without also considering air quality. We couldn't think about air quality without considering energy. And we couldn't think about energy, or any aspect of a city's future, without thinking about climate change. All of these issues were intertwined and interdependent.

A land use plan was clearly insufficient. We needed a comprehensive strategic plan that could guide the city's overall development in a way that would improve lives while also fighting climate change and protecting the public against its possible effects. Our goal was to allow the city to reap all of the benefits of population growth while avoiding its negative consequences.

Our administration had already done some work to reduce emissions. We had adopted one of the nation's first green building laws, requiring that all new buildings and capital projects receiving city funding meet stringent energy-efficiency standards. The Solid Waste Management Plan we created also drove down emissions—by shifting trash removal from trucks to rail and shipping. And while the United States had declined to ratify the 1997 Kyoto Protocol, New York had joined hundreds of cities around the country in pledging to meet and exceed U.S. carbon emission targets outlined in the agreement, which ultimately called for a reduction of at least 7 percent below 1990 levels. But, it was fair to ask, how would we do that while our population grew by more than 10 percent?

We began looking around at what some of the world's most ambitious leaders on climate change were doing.

In September of 2006, I flew to California to tour a fuel cell factory with then-governor Arnold Schwarzenegger. Arnold and I were both Republicans (I'm now an Independent) committed to reaching across the aisle to address challenges facing our nation, including climate change. The fuel cell company we toured was one of a growing number of tech firms that were strengthening California's economy while simultaneously shrinking its carbon footprint. Arnold had helped make California a leader by adopting an ambitious plan to cut state greenhouse gases by 25 percent by the year 2020. If California could do it, why not New York City?

To help guide our work, we created a new agency, the Office of Long-Term Planning and Sustainability, and a public-private advisory board composed of leaders from various industries and environmental groups. Three months later, our administration adopted ten goals that would guide our work in drawing up a plan for New York's future. They included giving New York City the cleanest air of any major city in America, and shrinking our carbon footprint 30 percent by 2030—two deeply ambitious goals.

How would we do it? At first we weren't sure. So we held dozens of public forums and town hall meetings, launched an interactive website, and held discussions with more than 150 advocacy groups to gather ideas and insights. We also brought on scientific advisers by forming a partnership with Columbia University's Earth Institute.

After gathering and analyzing a wide range of ideas, we launched

PlaNYC (called "plan-y-c") at the city's American Museum of Natural History on Earth Day 2007. Standing underneath the giant blue whale that hangs in the museum's Hall of Ocean Life, I outlined a plan that included 127 initiatives aimed at creating the world's most environmentally sustainable city. I asked the audience:

*If we don't act now, when?*
*And if we don't act, who will?*

It was an exciting moment but also a daunting one. We had just committed ourselves to an urban planning effort that was unprecedented in its scope. The next question was: Where do we start?

There's a saying I often used in city hall, and still use at my company and foundation: "If you can't measure it, you can't manage it." That's true in business, government, and philanthropy—and climate change is a perfect example. The only way to know whether the fight against climate change is succeeding is to quantify emissions, categorize them by source, and—after new policies are put in place—track them. An initial phase of PlaNYC was therefore to conduct a complete inventory of New York City's greenhouse gas emissions—giving us, for the first time, a picture of the total carbon footprint of everyone who lives in, works in, and visits New York City. This would enable us to identify the largest sources of emissions, so we could target them for attack. It would also allow us to track our progress and empower the public to hold us accountable.

The results of the inventory surprised us. Given the heavy traffic in New York City—the term "gridlock" was coined by a New York City traffic engineer in the 1970s—one might think that most carbon emissions came from cars, trucks, and buses. But in fact, because so many people walk and ride the subway—I rode it to work most mornings—we learned that about 75 percent of our emissions came from buildings.

Reducing those emissions meant examining every aspect of how the city worked, and how it could work better. We began looking at every possible option, and very early on, we recognized a central truth that the national debate over climate change got wrong: What was good for people and job growth was good for fighting climate change. Trees and parks give people opportunities for recreation and relaxation, and they also suck carbon and soot out of the air. Strong mass transit connects people to jobs and opportunities, and also reduces traffic and air pollution. Bike lanes connect neighborhoods and help improve public health, and they also help keep cars off the streets by giving people a safe alternative. Energy-efficiency measures save consumers money and clean the air while also shrinking the city's carbon footprint. *Most of the things that make cities better, cleaner, healthier, and more economically productive places also reduce carbon emissions.*

As we developed PlaNYC, we weren't shy about stealing good ideas from other places. Cities share many common challenges, and they don't have to reinvent the wheel. What works in one city may not work in another, but it's a good place to start—and being able to point to results in other cities makes it easier to win support for a new idea, especially when there is data to back it up. By talking with other mayors, we were able to build on their successes. South American cities like Bogotá and Curitiba had found smart ways to make public bus service faster, helping to take cars off the road. We learned from that work as we designed and rolled out New York's first bus rapid transit routes, which offer express service in dedicated lanes. In Europe, Copenhagen and Paris had used innovative policies to encourage more biking, like protected bike lanes and public bike sharing, while Berlin had found smart ways to encourage green roofs.

We also studied an idea that had shown promise in London, Singapore, and Stockholm: a toll on drivers that enter the center of the city during peak hours, to reduce traffic congestion, encourage the use of mass transit, clean the air people breathe, and reduce greenhouse gas emissions. Traffic congestion also harms the economy by slowing deliveries, raising prices for consumers and businesses. It was estimated that traffic jams cost the New York metropolitan region around $13 billion each year in higher prices and lost productivity.

Weeks before the PlaNYC launch on Earth Day 2007, city hall staff briefed me on their analysis of this idea—congestion pricing—and their proposal for implementing it in New York. I wasn't convinced it could work—the island of Manhattan has so many points of entry, through so many different neighborhoods, and businesses and residences all mixed together, making such a system unusually complex. So I sent them back to the drawing board.

This process repeated itself a few times, right up until the eve of the speech. I knew that political opposition would be stiff, but I took a careful look at the data and the potential benefits to the city and decided it was worth a shot. After all, the question was not whether we wanted to pay, but how: With rising asthma rates and ever-greater carbon emissions? With lost business and higher prices for consumers? Or by charging a modest fee to drivers, and then using that money to upgrade and expand subway and bus service? According to our team's estimates, congestion pricing would reduce rush hour traffic by more than 6 percent and increase the flow of traffic by more than 7 percent. That may not sound like a lot, but when you're the one sitting in traffic, whether you're running errands or running late for work, every second matters. As a bonus, the plan would also bring in close to half a billion dollars in revenue each year to invest in mass transit. I signed off on the proposal and we put it in the speech.

After we announced it, polling data showed that a strong majority of New Yorkers, when they were told that the money would be invested in mass transit, supported the idea. By being honest about the costs and benefits, we were able to build a coalition of supporters that included leaders of labor unions and business, and liberals and conservatives. The three major daily papers covering city politics—the *New York Times*, the *Post*, and the *Daily News*—all supported it, and they can barely agree on the time of day. The city council voted to approve it. But in New York, as in many cities, state approval is required to adopt certain laws. And our state legislature, which had recently been declared the most dysfunctional in the nation, was often where good ideas went to die.

After more than a year of spirited public discussion, with countless public forums, meetings, and debates, the leader of the New York State Assembly (later convicted on unrelated corruption charges) announced that our congestion pricing bill would not even get a vote. So much for democracy. In the years that followed, when people would complain about the lack of funding for mass transit, I'd say, "Gee, I wish someone would come up with an idea that would generate revenue for mass transit, clean the air, and reduce traffic congestion. Wouldn't that be great?" Congestion pricing makes too much sense for politicians to ignore it forever, but action takes time—and requires sustained public support.

Fortunately, we had more success in Albany on other PlaNYC initiatives. In 2008, we worked with the state to pass a tax credit for installing green roofs, which can help reduce air pollution and carbon emissions by insulating buildings. We also won state approval to offer a tax credit to buildings that installed solar panels. We persuaded the state to give the city authority to oversee cleanup of lightly polluted, often abandoned sites called brownfields. With that authority, we created a new Office of Environmental Remediation that dramatically sped the cleanup process, allowing us to develop new parks and affordable housing in neighborhoods around the city. We also worked with the state to strengthen standards for heating oil to make it less polluting, which helped us to clean the city's air and protect public health.

While we appreciated the state's support on these and other initiatives, regional and national legislatures often have political interests that cities don't share. Giving cities more authority to take action on their own—particularly on energy and transportation—is one of the most important steps we can take to address climate change. Instead, some state officials seem intent on ignoring the problem. During Rick Scott's time as governor of Florida, officials in the state's Department of Environmental Protection said that they were told not to use the terms "climate change," "global warming," and "sustainability." A state of denial can be a city's worst enemy.

Why isn't the critical role that cities can play in fighting climate change more widely recognized? Because, for a long time, environmental leaders were focused on a top-down approach to cooling the planet: some kind of international treaty committing the world's nations to reducing their carbon emissions. The effort to create such a treaty began at the UN's Earth Summit in Rio de Janeiro in 1992. One of the results of the Rio summit was the United Nations Framework Convention on Climate Change. Its founding objective was to stabilize greenhouse gas emissions, and each year since 1995, the UN has convened an international summit (technically a "Conference of the Parties," or COP) to assess progress. For two decades, the news wasn't good: Global greenhouse gas emissions continued to soar to new heights.

Negotiations over a global treaty on climate change long suffered from the same faults that doom other kinds of international treaties: Nations are hesitant to agree to any restrictions until everyone else has agreed to them. When one player refuses to agree to the rules, all the others have an excuse not to, either. It's like being in school. If one kid acts out, others start to as well—especially if there are no consequences. When the United States refused to ratify the 1997 Kyoto Protocol, which theoretically bound nations to targets for reducing their emissions, other nations could say: Why should we reduce emissions when the largest emitter of carbon isn't?

That was the question long asked by many nations in the developing world. But cities had a ready answer: Because it's in our interest to do so. After Washington refused to act, hundreds of U.S. cities, including New York, agreed to meet the Kyoto targets. Then in 2005, Mayor Ken Livingstone of London brought together representatives from eighteen cities to highlight their climate work and to help them share strategies. For the first time, mayors had begun thinking of cities not just as bit players in the climate discussion but as a collective force for change.

Within a year, eighteen cities had become forty, and the C40 Cities Climate Leadership Group was born. A month after we launched PlaNYC in 2007, New York City hosted a summit for the group, which today includes ninety cities that produce more than one-quarter of global GDP. (I serve as president of its board of directors.) Thanks in no small part to the recruiting efforts of former C40 Chair Eduardo Paes, who as mayor of Rio de Janeiro led an ambitious climate agenda, more than half of C40 cities now hail from outside of North America and Europe.

Later in 2007, I attended the UN's Climate Change Conference in Bali to highlight the work of cities and to make the case for why they should be given a bigger role in negotiations. City efforts were starting to gain recognition, but mayors were invited to Bali as part of national delegations, not as full participants with a voice of their own. In 2009, the UN held another climate summit, this time in Copenhagen, which I also attended. Hopes were high, after a U.S. presidential election in which both candidates promised to pass some form of cap-and-trade legislation, that the U.S. would join the rest of the industrial world in a new, improved version of the Kyoto Protocol. But the timing was bad, because the world was still feeling the hangover of a major global recession.

Fortunately, momentum was building in other ways. Major new investments in renewable energy made by the United States, Europe, and China were starting to pay off in a big way. Wind and solar in particular were becoming not only cleaner than fossil fuels but also cheaper. At the same time, more and more cities had evidence to show the economic benefits of fighting climate change. In New York, the investments we made hadn't just improved the quality of our air—they were leading to new jobs and businesses all across the city, and saving taxpayer money. *While nations argued about how to share the responsibility for taking on climate change, cities were increasingly taking the responsibility upon themselves—both independently, and in cooperation with other cities.*

After Copenhagen, negotiators realized that instead of a top-down climate agreement like Kyoto, they could construct a bottom-up solution defined by commitments from a variety of parties—cities, businesses, industries, and nations. The UN's new head of climate negotiations, Costa Rica's Christiana Figueres, underscored the point, declaring, "This planet is not going to be saved by any big bang agreement." For the first time, the UN embraced a philosophy that cities had long ago adopted: think globally, act locally.

In 2014, UN Secretary-General Ban Ki-moon asked me to serve as Special Envoy for Cities and Climate Change. I accepted, and focused on encouraging cities to adopt bold climate agendas and helping to bring more transparency and accountability to their climate work. A number of efforts were under way to help more cities measure and report their carbon emissions, as we had done in New York, but a single, transparent reporting standard had yet to be adopted. This made it difficult for cities to measure their progress against one another, and for nations to gauge their progress.

Through my work as special envoy, Bloomberg Philanthropies joined the UN and the European Commission to create an organization that would come to be called the Global Covenant of Mayors for Climate and Energy. Under the Global Covenant, cities commit to publicly measuring and reporting their carbon emissions using a standard measurement system. Today, our group includes more than 7,000 cities in 112 countries.

In December 2015, the UN's twenty-first climate conference met at a building in the suburbs of Paris, where heads of state and senior government officials gathered to negotiate an agreement. Meanwhile, in the heart of the city, at Paris's city hall, the Hôtel de Ville, representatives of nearly 500 cities from 115 countries descended upon the first-ever Climate Summit for Local Leaders, organized by Paris Mayor Anne Hidalgo in partnership with Bloomberg Philanthropies. And just across the Seine, outside the Place du Panthéon, artist Olafur Eliasson and geologist Minik Rosing had installed twelve blocks of glacial ice, arranged in the shape of a clock. As negotiators deliberated a new agreement, the blocks slowly melted.

That sense of urgency was echoed by the mayors gathered in the Hôtel de Ville. The city summit was an old-fashioned show of force, intended to send a clear signal to national leaders that mayors were committed to implementing bold climate change policies and demanding that they reach an ambitious agreement. The strategy worked. By flexing our muscles, we earned a voice in the process that we had never had before. The Paris Agreement committed nations to hitting emission reduction targets, and—thanks to the example set by mayors—it included a reporting provision, allowing the world to track each nation's progress. "Our sheer numbers helped influence the language put into the deal," said George Heartwell, mayor of Grand Rapids, Michigan. "Success of the international agreement requires the cooperation of us, the cities," he continued. "I left Paris feeling inspired. If Grand Rapids can [do] this, every city can also."

The Paris Agreement was not achieved through a newfound altruism on the part of nations. It was driven by the realization that cutting emissions held economic and health benefits for nations that they had previously undervalued. Cities had demonstrated that these benefits were real, which encouraged nations to set high goals. Paris was a big step in the right direction, and it paved the way for the international aviation industry to adopt a new market-based mechanism to limit emissions, and for the world to agree, in 2016, on a timetable for phasing out the heat-trapping hydrofluorocarbons (HFCs) that still power many air conditioners and refrigerators.

After the 2016 U.S. presidential election, there was a lot of speculation about whether the United States would fulfill the pledges our nation made in Paris. On the campaign trail, Donald Trump had vowed to "cancel" U.S. participation in the agreement. Before he became a presidential candidate, Trump had once gone as far as to say that climate change was a hoax invented by the Chinese to reduce U.S. competitiveness. But, as I said at an event hosted by the China General Chamber of Commerce two weeks after the 2016 election, no matter what happens in Washington, no matter what regulations the Trump administration adopts or rescinds, no matter what laws Congress may pass, market forces, local (and in some cases state) governments, and consumer demand for cleaner air will, together, allow the United States to meet and exceed the pledges that the Obama administration made in Paris.

The reason is simple: Cities, businesses, and citizens will continue reducing emissions, because they have concluded—just as China has—that doing so is in their own self-interest.

If this seems overly optimistic, bear in mind that over the past decade Congress has not passed a single bill that takes direct aim at climate change. Yet at the same time, the United States has led the world in reducing emissions. That progress has been driven by cities, businesses, and citizens—and none of them intend to let up now. Just the opposite: all are looking for ways to expand their efforts.

*America's ability to meet our Paris climate pledge doesn't depend on Washington.* It depends on cities continuing to protect their residents and invest in the future. It depends on businesses continuing to seek ways to save and make money. It depends on technology continuing to make renewable energy more affordable. And it depends on citizens continuing to demand cleaner forms of energy that won't harm their health and pollute their communities.

Stronger leadership from Washington would be greatly welcomed, but Washington will not have the last word on the fate of the Paris Agreement in the United States. Mayors will—together with business leaders and citizens from all over the globe. As the mayor of London, Sadiq Khan, says: "Climate change is one of the biggest, if not the biggest, risk to London. I want to put our city at the leading edge of the fight against this risk. But London can't take on this challenge alone. That's why I have joined forces with leaders of world cities to find ways that we can create cleaner, greener, better, and healthier cities together."

Climate change may be the first global problem whose solution will depend on how municipal services such as energy, water, and transportation are delivered to citizens. Cities around the world have only just begun to seize the opportunities available to them to make changes that will yield both local and global benefits. Too often, however, they are stuck in the same position New York City was stuck in with congestion pricing: dependent upon another level of government for approval.

Countries must do more to empower their cities to create cleaner infrastructure. Take energy. Mayors in some cities—including Chicago, Seattle, Helsinki, and Toronto—enjoy various forms of leverage over their energy supplies. Some own their own power, others own the distribution system, and still others have the authority to sign contracts with any independent power generator they select. The Chinese government has given major cities, such as Shenzhen, expanded powers to swap out coal for cleaner forms of energy. In Denmark, the national government decided to grant independent regulatory powers to Copenhagen, which is now aiming to reach carbon neutrality, or zero net emissions, by 2025.

Thanks to PlaNYC, by the time we left office at the end of 2013, New York City's carbon footprint had shrunk by 19 percent—putting us ahead of schedule to reach a 30 percent reduction by 2030, even as our population continued to grow. At the same time, air pollution fell to the lowest levels in a half century, and New York City created a record number of jobs. With more autonomy, however, we could have made even more progress.

Central governments are not quick to devolve power, but they are doing so with greater frequency as they recognize the national benefits that can come with local control. That trend will only accelerate as the world becomes increasingly urbanized and cities become increasingly connected to one another, promoting the spread of best practices across national borders. City leaders don't seek to displace their national counterparts but rather to be full partners in their work. And the faster nations embrace their cities as partners, the faster we can make progress on climate change.

I'm optimistic that this trend will accelerate, but not because Congress will become more enlightened. Congress does not lead; it follows. To the extent that the public see things changing, I think you'll see Congress changing: slowly but surely. The good news is that it's already starting to happen, as voters all over the country see storms growing stronger and more frequent, as they see floods where they never had them before, and as they suffer through droughts that are worse than they've ever experienced. Americans are a lot smarter than the elected officials they send to Washington. Our country's citizens want to avoid these disasters—and they know they can do something about it. They want to breathe clean air—and they know they can do something about it. They want to reduce their expenses—and they know they can do something about it. That's why mayors—who are most responsive and accountable to the public—are taking action. Mayors have gotten the message, and eventually national legislators will, too.

# PART II

# WHAT IT IS AND WHY IT MATTERS

3

# THE SCIENCE

*It's really quite simple. We've overloaded the atmosphere
with heat-trapping gas and the rest are just details.*
—Professor Jason Box, Geological Survey of
Denmark and Greenland

I first learned that global warming was an immediate scientific phenomenon in 1987. A former colleague, Rafe Pomerance, came to see me at the Sierra Club. He explained that the abstract notion initially advanced in 1896 by Swedish chemist Svante Arrhenius—that carbon dioxide accumulating in the atmosphere from burning fossil fuels would disrupt the planetary climate—was soon going to be a tangible and dangerous fact of life, frighteningly evident in the world around us.

My initial reactions were: "We're going to have to explain this very carefully to avoid a public panic," and "This is very bad news for the coal industry." The first was laughably off base; the second has proven true, but only after a very risky lag time. I completely underestimated the denial and resistance that the scientific evidence for climate change would provoke. The evidence pointed to more rapid change than had been expected. Rather than triggering action, it led to a quarter-century-long ideological war on science by fossil fuel interests and ideological opponents of government action.

Nor did I understand that, while coal is, indeed, the largest single source of climate pollution, an entire collection of pollutants was actually driving the problem.

## WHAT'S CLIMATE CHANGE, ANYWAY?

When sunlight—solar radiation—hits the Earth, a portion of it is absorbed as heat, and the rest is reflected back into space. Darker objects—forests, soils, and, most important, the oceans—absorb a larger share of the energy in sunlight, while lighter soils, clouds, and ice caps absorb less and reflect more back into space.

Different gases and particles in the atmosphere capture some of the solar energy both as it hits the atmosphere and as it bounces back as heat. These are the greenhouse gases, named after the way in which a greenhouse keeps solar energy in and heats up the temperature. (Without such gases, the Earth would be like the moon—which is 253 degrees Fahrenheit by day and plummets to 243 degrees below zero at night.) Water vapor captures the lion's share (about 60 percent); carbon dioxide is the second-largest heat-storage gas in the atmosphere; and methane, or natural gas, the third. This stored sunlight in the Earth's atmosphere drives the entire weather cycle: winds, the creation of clouds, hurricanes, heat waves.

Svante Arrhenius's 1896 theory was quite simple. Burning fossil fuels (and, it turns out, cutting down forests and degrading soils) increases the amount of carbon dioxide in the atmosphere. This causes the blanket of air surrounding the Earth to retain more solar energy. More retained solar energy has two effects. It causes temperatures to rise—hence global warming. It also makes the atmosphere more energetic, just as heating a pot of spaghetti sauce not only heats it but eventually causes it to bubble over. Storing more—and reflecting less—of the incoming sunlight puts weather on steroids.

Arrhenius didn't know it, but industry, mining, and agricultural practices were changing the chemistry of the atmosphere in other ways: increasing concentrations of methane by extracting coal, oil, and natural gas; keeping ever-larger herds of livestock, which belch and fart methane (as do humans, in smaller quantities). By the mid-twentieth century, industry was putting a variety of additional heat-storing chemicals into the atmosphere as well, some of which—like the halocarbons used as refrigerants and in air conditioners—are extraordinarily effective scavengers and retainers of solar energy.

Climate change is not in question. The overwhelming consensus of scientists who study the atmosphere and climate is that current levels of man-made greenhouse pollutants are large enough to produce very serious disruption of the climate, with potentially catastrophic impacts, and that these effects can already be measured. There are dissenters (climate skeptics), but they are a tiny minority with virtually no significant peer-reviewed science to back them up. Emitting enough of these "greenhouses gases" changes the chemistry of the atmosphere and the patterns of the weather.

Weather is a "chaotic" phenomenon. A small change today can result in a large change tomorrow. And climate is a linked system, one where changes produce numerous reactions. Warmer air means more water vapor evaporates from the oceans, which can in turn produce more snowfall in Canada but also stronger hurricanes in the Caribbean. A warmer atmosphere therefore inevitably means altering the climate of the planet in complex and difficult-to-predict ways.

Even scientists who call themselves "climate skeptics" don't really argue this basic theory. They concede that changes in the chemistry of the atmosphere will alter the climate and the weather it generates. They simply claim that we don't know how big such effects will be, when they will occur, what we can do about them, or how much it matters. Some of them argue that while average and regional global temperatures have increased measurably—almost a degree Fahrenheit in the last half century—some, and perhaps all, of this rise might be due to preexisting trends toward a slightly warmer world unrelated to man-made emissions.

More troubling, climate skeptics don't just minimize the likelihood and severity of climate changes produced by greenhouse gases; they also don't agree that we ought to take action to reduce the risks of such changes, even though they do agree that we are—unavoidably—uncertain of exactly how big they are or how quickly they might arrive.

But, of course, I buy fire insurance even though my house might never burn down. I carry health insurance even in years in which I don't get sick. Businesses hire security guards although they might not be robbed. Airport security pats down implausible terrorists like seventy-year-old environmentalists (among them me). The normal human response to uncertainty and risk is to take reasonable precautions. The uncertainty of climate model projections means that things are equally likely to be worse or better than we expect—and worse could mean very bad things.

My "Aha!" moment—the moment when I grasped how big a change in climate might be in store for us—came in February 2011, when I took a surreal walk through New York's Central Park. A bedraggled sign warned of "Thin ice—Danger" at one of the ponds. It was 72 degrees, so any ice was very thin indeed. A hundred yards ahead, a Zamboni machine at the park's ice rink strove earnestly to freeze up a skateable surface, all the while pushing a bow wave of melted water ahead of it, like an amphibious landing craft hitting the beach.

The Zamboni's plight struck me as a good metaphor for what humanity faces. Climate has always been bigger than technology; our winning strategy has centered on our flexibility. We have learned to live within the fixed cycles and normal weather patterns in diverse places. We've adapted both to the freezing cold of Baffin Island and to the scorching heat of the Kalahari Desert. We've proven our ability to thrive in a wide variety of climate patterns. Underscore *patterns*.

Climate simply means predictable patterns of weather. At least it did.

Now industrial mastery and scale—extracting and burning fossil fuels, cutting forests, creating voracious heat-retaining chemicals—is making the climate erratic. Global warming is destroying the very thing that has made us successful in coping with climate—patterns and predictability. And not knowing what is going to happen should scare us. No ice-skating in Central Park is the least of our worries.

## HOW 12,000 YEARS OF STABLE CLIMATE MADE CIVILIZATION POSSIBLE

One of the least understood pieces of the climate puzzle is that human civilization emerged in, and was nurtured by, an unusually favorable climatic moment. It's true, as climate change deniers point out, that climate has often changed in the past, sometimes very dramatically. When you consider that most of the northern United States was under a mile of glacial ice only 18,000 years ago, ice that took 6,000 years to retreat; that Florida, now under risk from a few feet of sea level rise, was once almost twice as wide when sea levels were lower; and that Alaska was once connected to Russia by land, it's easy to say, "Well, what is the big deal? Climate is always changing." Well, yes, but not necessarily to the advantage of human beings, and not always at the same pace. During most of the Earth's history, it would have been very hard for *Homo sapiens* to journey from the Stone Age to the iPhone. The climate was either too extreme (that wall of ice) or changing too rapidly for settled communities to adapt.

As the glaciers gave up their hold on North America, a series of glacial melt lakes burst their shorelines, precipitating a wave of floods; had there been urban civilization along the coasts during this period, it would have been wiped out. The last of these glacial mega-impoundments, Lake Ojibway, burst out northward into Hudson Bay, releasing so much water that it raised sea levels dramatically on the Gulf Coast and globally.

Once most of the glaciers were gone, however, the Earth's climate settled into an unusually favorable and stable system called the Holocene period: Almost 12,000 years of weather in which cities like Jericho could survive for thousands of years in the same location. This climatic stability enabled hunter-gatherers to figure out the tricks of raising crops and domesticating animals. Paddy rice cultivation, for example, was perfected in coastal areas of eastern China—but only because sea levels remained relatively constant.

Getting used to a new climate takes time. Even the Holocene has not been immune from climate ups and downs. The retreat of the remaining glaciers has dried global climate, creating the hostile deserts of Africa, Arabia, and Central Asia, where cities and trade routes once flourished. Medieval settlements that were once part of today's Belgium now lie under the North Sea. Viking settlements in Greenland, built during a temporary medieval warming period, were starved out when colder temperatures returned. They couldn't adapt. And the gradual drying out of Central Asia unleashed the vast nomadic migrations that led first to the barbarian assaults on the Roman Empire and later to the Mongol occupation of China, Russia, and the Middle East.

So the last 12,000 years blessed us with a climate stable enough—barely—for civilization to emerge and thrive. But rapid climate change denies communities the time they need to adapt. And as civilization has become more settled and complex, it has also become more vulnerable to small changes in the weather. Nomads can shift with the rains—even over long distances. An Iowa farmer facing an extended drought cannot. Cities are even less mobile, and changes in climate bring with them unanticipated challenges. A small example: Homes in Cincinnati are protected against termites; termites were always there. Homes in Chicago are not—winters were too cold. Given what climate scientists can tell us, should Chicago undertake an expensive and massive termite-proofing exercise for its frame buildings? When? How fast? Not knowing is risky indeed for complicated, sedentary societies.

# THE JAZZ ENSEMBLE OF POLLUTANTS THAT THREATENS THE CLIMATE

The debate about climate change has mostly been about whether we should move beyond fossil fuels for energy. But the problem is far larger than fossil fuels, which are only one source of $CO_2$.

$CO_2$ also results when iron ore is converted to steel, or limestone to cement, or oil into chemicals. Such emissions amount to about 20 percent of $CO_2$. Another 15 percent of $CO_2$ emissions come from deforestation. A few years ago, before Brazil began protecting the Amazon much more effectively, that proportion was 20 percent. And poor agricultural practices turn soils, which normally store carbon, into sources, releasing $CO_2$ into the atmosphere.

Moreover, carbon dioxide is only one source of the problem. There is a whole group of gases and particles that grab on to solar heat and keep it from radiating back into space. Fossil fuels emit greenhouse pollutants other than $CO_2$. Incomplete combustion of fossil fuels—whether the result of dirty diesel fuels in ships and trucks or wood and biomass (often used to cook in developing countries)—emits soot, which scientists call "black carbon." Soot is a major greenhouse pollutant because, being black, it absorbs and holds solar radiation. It is particularly powerful when it is released near glaciers or ice caps, because it smudges otherwise highly reflective ice surfaces and causes rapid glacial melt. So black carbon is, for now, the second largest force disrupting the climate.

After $CO_2$ and black carbon, the third most damaging climate pollutant is methane, a natural gas. Methane has been created for hundreds of millions of years when plants and animals decay in the absence of oxygen in wet environments (like swamps). Methane is still being created in this way, and about a third of current methane emissions come from today's swamps, bogs, and forests. The other two-thirds result from human activities. As a result, methane concentrations in the atmosphere have soared even faster than $CO_2$.

A lot of this additional methane comes from waste and agriculture.

Sewage and garbage sent to landfills decay and produce 11 percent of annual methane emissions. Rice paddies, which work like swamps, generate another 12 percent. Cows and other livestock, through their digestive bacteria and rotting manure, produce 21 percent.

But more and more of the methane being released into today's atmosphere was made a long time ago, as part of the process that turned Carboniferous age forests into coal, oil, and natural gas. As we extract fossil fuels from the Earth, we release a lot of methane along with them. Mining and pulverizing coal releases 8 percent of global methane. Oil and natural gas wells and pipelines carrying that gas to your furnace, stove, or a neighboring power plant yield another 12 percent.

Per molecule, the most powerful heat-trapping molecules we know of are the recently invented halocarbons—industrial chemicals that combine carbon with halogens: chlorine, fluorine and bromine. The biggest threat is from hydrofluorocarbons (HFCs), fast becoming the standard refrigerant in air conditioners and cooling systems. HFCs, ironically, replaced chlorofluorocarbons (CFCs), which were destroying the ozone layer.

Some of the halocarbons are 7,000 times more effective at storing heat than $CO_2$. So even though we emit a relatively small volume each year, halocarbons have accounted for 17 percent of global warming to date. Fortunately, in 2016 in Kigali, the world community agreed to a phase-out schedule that will eliminate much of the climate risk from these refrigerants, replacing them with safe and affordable alternatives.

## THE LIFE CYCLE OF CLIMATE POLLUTANTS

These varied climate pollutants have different fates once emitted. Soot (black carbon) from forest fires, open cooking, and diesel engines falls out of the atmosphere within six months of being emitted. A molecule of methane holds about 84 times as much heat as a molecule of $CO_2$, but in the atmosphere it encounters various compounds that break it apart, so its average lifespan once emitted is only twelve years. CO2, the pollutant scientists worry about most, lasts in the atmosphere for about a thousand years—unless it is absorbed by the ocean, or taken up by a greedy young plant looking to combine it with sunlight to make a leaf. Halocarbons used as refrigerants are most effective at holding solar heat, and some last thousands of years before breaking down.

So, as you are reading this book, any black carbon produced by a dirty diesel engine before Mike and I starting writing this book fell out of the atmosphere as dust long ago. The only methane still around was released after Al Gore wrote *An Inconvenient Truth*, even though the amount in the atmosphere has increased significantly since that book. Essentially all of the halocarbons ever emitted are still grabbing and holding onto heat. And the CO2 from burning fossil fuels for the last 1,000 years is still around, too, but about 60 percent of it has been absorbed by oceans or forests and soils.

Since we can't be sure how many of these gases we will emit in the future, or how they will interact with one another and the weather, precise projecting of future climate is impossible. But this uncertainty should galvanize us to tackle its various causes, rather than be used as an excuse for inaction.

## UNCERTAINTY AND CLIMATE CHANGE DENIAL

It has always been hard to predict the future: economically, politically, and meteorologically. But it is scandalous that this inability is now being used as an excuse to do nothing about the greatest threat facing human life on our planet today.

During my years debating climate science as the head of the Sierra Club, I was perpetually frustrated by the media obsession with which specific weather events we could safely attribute to climate change.

Something big and dramatic—a hurricane, a blizzard, a drought—would devastate communities, revealing starkly just how dependent on a friendly climate we are. But instead of heeding these warnings that we shouldn't keep running a global chemistry experiment to see how wild we can make the weather, the media would debate whether this particular tragedy could be conclusively attributed to climate change.

In one typical example, the *New York Times* proclaimed in 2013 that the inability to answer these questions was a failure of climate science: "The new report also reiterates a core difficulty that has plagued climate science for decades: While averages for such measures as temperature can be predicted with some confidence on a global scale, the coming changes still cannot be forecast reliably on a local scale." A *Times* blogger piled on, saying, "The reality is that efforts to attribute shifts in patterns of extreme weather to the greenhouse effect remain dogged by enormous uncertainty."

I would frame it very differently. I would argue that our inability to predict with specificity the new climate normal for a particular region is not a failure of climate science—it is a warning about how dangerous climate disruption can be.

After all, when critics called on Major League Baseball to deny recognition of Barry Bonds's home run records because of his alleged steroid use, they did not attempt to calibrate *which* of his four-baggers might have resulted from steroid use, or even to determine what his batting average without steroids would have been. They understood that steroids are unacceptable because they change the odds, not because we know what they do in any given at bat by any particular player.

Climate is not a single event; it is a statistical summary of weather over time. It is the batting average as opposed to an at bat. And the concept of "global climate" requires adding up the individual weather that occurs over decades in thousands of locations. That is why it has been so hard to document how global "climate" has already changed.

It is much simpler to understand the risks—and let's shift to a kitchen metaphor by returning to that pot of spaghetti sauce I mentioned earlier. Imagine it sitting on a stove. The burner is off. Not difficult to predict what the sauce will do—it will sit. Now imagine the burner is on simmer. The sauce bubbles gently, but you still don't need to worry about anything dramatic. That's the climate the world has enjoyed for the past 12,000 years, the Holocene. In spite of individual tornadoes, floods, cyclones, and droughts, human civilization developed in diverse locations. There was time for people (and other species) to figure out how to make a home in a given place, because the weather didn't change that much. Chicago might have its cold and warm seasons, but it was not going to have Riyadh summers some years.

Now turn up the burner, steadily higher, making the sauce hotter.

At some point the spaghetti sauce will start splattering, and eventually boil over. As you keep turning the burner ever higher, you will have a very hard time guessing exactly when and which way the sauce will fly. But splatter it will.

The fossil fuel apologists have latched on to this search for predictability to undermine the ample scientific evidence that our climate is at risk: "See," they say, "they don't even know if it is going to be drier or wetter in Iowa. How can we believe science that is so imprecise?" But that's the point: The increased uncertainty is a byproduct of our already changing climate.

One major argument by the skeptics is that perhaps cloud cover will magically make up for the warming impact of other climate pollutants. Water vapor is the biggest storage sink for heat in the atmosphere. More water vapor does mean more clouds, and clouds do reflect sunlight back out to space because they are white. But while cloud cover will reduce warming, there is no reason to believe that the net effect of increased water vapor will automatically stabilize the climate—the geologic record shows that in the past, when $CO_2$ concentrations went up, the climate was disrupted in major ways.

As we alter what the atmosphere is made of, and therefore how much energy it stores, weather—and climate, which is simply the fixed patterns of weather—becomes less predictable. We cannot say today, and may never be able to say, what the weather in Des Moines will be next August 20. There is therefore no particular reason to believe that climate models will be able to tell us with much accuracy what the summer weather will be like in New York in fifty years if we triple concentrations of $CO_2$ over that period but cut methane by 25 percent.

That imprecision, of course, is exactly what we should expect with a chaotic phenomenon. Nevertheless, weather has patterns—attractors, mathematicians call them. We don't know how hot Des Moines will be in a year, but we can say what the average temperature and rainfall are likely to be. Those patterns—the climate—underlie chaotic weather.

What science and experience are already telling us is that if we keep adding greenhouse pollutants to the atmosphere, we should expect more extreme and violent weather. Indeed, once we disrupt the current patterns, we really shouldn't expect to know what the new climate will be.

This mania for certainty where we are unlikely to find it has become a perilous distraction. Would you really keep turning the burner up under a pot as it boiled over because you weren't certain how much time you had to cool it? Or legitimize steroid use in baseball until you could calculate just how big an edge steroids gave a batter? If we were simply dealing with a pot of spaghetti sauce, the solution would be self-evident—turn down the burner. But industrial civilization is much more complex.

Rather than thinking of climate as one problem, caused by burning fossil fuels for energy, we need to understand it as a group of problems. Coal mining, steel and cement production, forestry and agriculture, oil and gas extraction, and refrigeration and cooling are each altering the chemistry of the atmosphere in a different way. The combination of these alterations has put the stability of the climate in peril. Replacing fossil fuels with clean energy is an excellent place to begin fighting climate change. But it isn't an adequate plan to restore a stable climate.

## ONE BATTLE, MANY FRONTS

It may be surprising to learn that *Homo sapiens* began tweaking the weather long, long before we burned the first lump of coal. For millennia, hunter-gatherers burned forests; later, in Europe and Asia, the need for timber, and the clearing of lands for agriculture along the floodplains of rivers, dramatically reduced forest cover and released $CO_2$. Overgrazing and poor farming practices in the pre-modern era stripped much of the carbon from soils in the Mediterranean, China, and India. Until 1960, deforestation and farming were changing the climate faster than fossil fuels.

Since the Industrial Revolution, the buildings in which we live and work have become a major contributor to global warming. If we include the electricity and the fuels we use in our homes and offices—for lighting, computers, appliances, and elevators, but particularly heating and cooling—buildings are responsible for about a third of climate change.

Transportation has also become a significant contributor to the problem, given our overwhelming dependence on cars, trucks, and planes—and the fact that more than 90 percent of transportation is powered by oil. Finding substitutes for oil in transportation remains more challenging than generating electricity or heating and cooling buildings without fossil fuels.

Our dependence on electricity is also vast. Edison's first power station in New York burned coal—so even the very first electrons used to light a bulb had a tiny climate impact. Then for decades hydroelectric power was the big source for electricity, and the impact on our atmosphere diminished. Over time, however, as good dam sites for hydro ran out, coal became the "go-to" fuel for electricity. As a result, the generation of electricity is at the center of the climate debate. All three fossil fuels— coal, oil, and natural gas—are major sources of electricity. All told, generating electrons drives about 25 percent of climate warming. But electricity is also the place we have made the most dramatic progress in getting rid of our need for fossil fuels, with both wind and solar in many places now providing electricity much more cheaply than fossil fuels like coal.

Finally, stuff we make—steel, aluminum, cement, plastic, toys, clothing, furniture, chemicals, vehicles—accounts for a solid 21 percent of total climate pollution. Two-thirds of that is carbon released by the steel, petrochemical, and cement industries as part of their core processes, and is thus particularly difficult to eliminate.

In other words, climate change isn't one problem with a single silver-bullet solution. The climate is being disrupted by a variety of pollutants, and these pollutants come from many different parts of our lives. They don't work alike; they have different lifetimes. Some are easy to replace, others harder.

Unfortunately, a lot of the climate debate has focused on the need for single solutions. Some think we primarily need a research-based "energy miracle." We do need to spend a lot more on clean energy research, but even if we got a miracle that solved the problem of $CO_2$ emissions from energy, it wouldn't solve climate pollution from cement kilns, methane from livestock, or halocarbons from refrigeration.

Another silver bullet was a political favorite from Kyoto to Copenhagen: cap and trade. Cap-and-trade programs set a quota on climate pollutants, then assign the right to emit these pollutants, and then allow those who want to emit more pollution to purchase from others willing to emit less. Politicians often like cap-and-trade systems because they can assert they are "giving" something away—the right to emit. But the European version has broken down repeatedly. In the United States, even when Democrats controlled the Congress, public skepticism killed the idea, although California is proving it can work.

Most economists think a carbon tax is simpler and better. It's a better way to fund a government than taxing productive work. The public support it, particularly when they know the revenues will be used to fund clean energy. But higher prices only change behavior when markets are competitive. When there is a monopoly, as there is for airplane travel, expensive jet fuel just means higher ticket prices; people don't stop flying. So no single solution is going to work in all sectors of the economy or all countries.

The good news behind this complexity is that it gives us not one— or even two or three—powerful avenues to reduce these emissions, but dozens. Virtually every sector of society in every country can make some significant contribution toward climate protection and be better off in the short term for doing so, and in the chapters ahead we'll explain how.

4

# THE STAKES

*Turning the Miami region into a real-world Atlantis*
*is a fate we cannot accept.*
—Cindy Lerner, former mayor of Pinecrest, Florida

I've had a passion for science for as long as I can remember. As a kid, I would travel on Saturdays from my house in Medford to the Boston Museum of Science, which sits on the Charles River near Boston Harbor. All of us remember the first teacher to have a profound impact on our lives. For me, that teacher was the Museum of Science.

My visits to the museum taught me to ask questions and sparked a lifelong interest in understanding how things work. That led me in high school to take a summer job at a small electronics company, where my supervisor encouraged me to apply to Johns Hopkins University. At Hopkins, I originally intended to get a degree in physics. There was, however, a German-language requirement, because many of the physicists whose work formed the basis of the curriculum were Germans. After three days in German class, I switched to electrical engineering. Although I ended up pursuing a career in business instead, I never lost my love of science and interest in figuring out how things work. And I have the Boston Museum of Science to thank for that.

In 2016, I returned to the museum to announce a gift in honor of my parents, Charlotte and William Bloomberg. I couldn't think of a more fitting way to honor my most influential teachers—my parents and the museum—than by linking them together and expanding the museum's educational work, so that more kids could have the same opportunities there that I had. Naturally, climate change will be one of the topics they learn about.

Climate change threatens life on the planet like nothing else. If, God forbid, we should ever have a nuclear war, most natural life on our planet would survive. Climate change, on the other hand, could do irreversible harm to the natural resources humanity depends on. If we do nothing, the long-term consequences to our planet could be cataclysmic. But as we've seen over the last two decades, it's counterproductive to focus on end-of-the-world scenarios. The reality is: warning of some far-off possible harm doesn't spur politicians to act, as they are motivated by short-term interests.

Journalists like to focus on these far-off dangers, too, because disaster makes for good copy. Economists also like to talk about the long-term costs of climate change, because the numbers are so extraordinary. Because of the uncertainty Carl described in the last chapter, efforts to calculate the total climate change bill are by definition speculative. But we know more than enough to be concerned. British Academy President Lord Nicholas Stern's 2006 *Review on the Economics of Climate Change* estimated that unchecked climate change would cost the world 5–20 percent of global GDP, but that the cost of confronting the problem would run about 1 percent. In the Risky Business Project, an initiative I have worked on with former Treasury Secretary Hank Paulson and former hedge fund executive Tom Steyer, we produced a report called *The Economic Risks of Climate Change.* The report documented that the costs to the United States would be very significant: In the decades ahead, coastal storm damage could grow to $35 billion annually, agriculture could face yield losses of more than 10 percent, and increasing power demand caused by rising temperatures could cost ratepayers an additional $12 billion annually.

*(Above) Sesame Street*'s Big Bird helps Bloomberg and Bette Midler launch Million Trees NYC in the South Bronx, 2007. *(Courtesy of the City of New York)*

*(Right)* Bloomberg with then California Governor Arnold Schwarzenegger, 2007.

*(Below)* Inside of a solar-powered vehicle during the UN Climate Summit in Bali, Indonesia, 2007. (*Courtesy of the City of New York*)

(Above) Addressing a climate conference at the United Nations in New York City. Secretary-General Ban Ki-moon seated on right, 2008. (Courtesy of the City of New York)

(Left) With Congresswoman Carolyn Maloney, City Council Speaker Christine C. Quinn, and others to announce expanded ferry service across New York, 2008. (Courtesy of the City of New York)

(Below) Painting a roof with former Vice President Al Gore during an event to launch NYC Cool Roofs, 2009. (Courtesy of the City of New York)

(Above) Looking north on Broadway in Times Square, a 2010 public art piece by artist Molly Dilworth, *Cool Water, Hot Island,* brought color to the pedestrian plaza. (*Courtesy of Molly Dilworth*)

(Right) Meeting Vancouver Mayor Gregor Robertson in New York's City Hall to discuss climate change, 2010. (*Courtesy of the City of New York*)

(Below) Presenting an update to PlaNYC, 2011. (*Courtesy of the City of New York*)

*(Top)* Announcing his commitment to the Sierra Club's Beyond Coal campaign in front of the GenOn Potomac River coal plant in Alexandria, Virginia, 2011. *(JEWEL SAMAD/AFP/Getty Images)*

*(Above)* Helping a child put on a helmet distributed through a Bloomberg Philanthropies road safety program in Vietnam, 2012.

*(Left)* Bloomberg is briefed in New York City's Office of Emergency Management as Hurricane Sandy approaches, 2012. *(Courtesy of the City of New York)*

*(Above)* Bloomberg and Deputy Mayor for Operations Cas Holloway survey damage after Hurricane Sandy in New York City, 2012. *(Courtesy of the City of New York)*

*(Right)* Bloomberg joins Transportation Commissioner Janette Sadik-Khan to launch Citi Bike, America's largest bike share program, 2013. *(Courtesy of the City of New York)*

*(Below)* With then C40 Chair Rio de Janeiro Mayor Eduardo Paes in South Africa for a C40 summit, 2014.

Climate Summit for Local Leaders organized at Paris City Hall by Bloomberg and Paris Mayor Anne Hidalgo during the UN's 2015 climate summit.

Speaking with Minister Xie Zhenhua, China's Chief Negotiator for Climate Change, during the Climate Summit for Local Leaders in Paris, France.

Speaking at the Climate Summit for Local Leaders; seated: Paris Mayor Anne Hidalgo and French President François Hollande.

*(Right)* Speaking with Patti Harris, CEO of Bloomberg Philanthropies and inventor and entrepreneur Elon Musk at the Climate Summit for Local Leaders.

*(Below)* Bloomberg Philanthropies supported artist Olafur Eliasson's public art project, *Ice Watch,* during the 2015 UN climate summit in Paris.

*(Right)* Bloomberg, Paris Mayor Anne Hidalgo, Mexico City Mayor Miguel Mancera, and Rio de Janeiro Mayor Eduardo Paes at the C40 Cities Climate Leadership Summit in Mexico City, Mexico, 2016.

*(Right)* Carl Pope retraced the steps of Sierra Club Founder John Muir to Glacier Bay from 1879. The Muir Glacier, named after him, has now retreated 31 miles, in a dramatic example of how vulnerable the world's glaciers are to climate change.

*(Below)* Pope meets with then Vice President Al Gore at the White House, 1996. *(Presidential Materials Division, National Archives and Records Administration)*

But when people emphasize the enormous challenges we'll face if we do nothing for the next one hundred years, it can make the problem of climate change seem too big to manage, which is discouraging—and false. Placing too much focus on these long-term risks also hides the most compelling reason to fight climate change, which is not the danger we face in the future but the deadly reality we already face today.

According to the World Health Organization, seven million people die from air pollution each year. That's about as many people as live in the cities of Houston, Chicago, Philadelphia, and San Francisco combined—all dying because of the air they breathe. That tragic toll makes air pollution one of the biggest risk factors for death globally, contributing to one in every eight premature deaths each year. And much of that pollution is caused by the same fossil fuels that are warming our planet—especially coal. Particulate matter from burning coal contributes to strokes, heart disease, lung disease, and cancer. If we could eliminate all the coal-fired power plants in China and India alone, we would save half a million lives every year. Of course, we can't completely make that transition this year or next, or maybe even in the next decade. But every step we take in that direction saves lives. Not in the future—right now.

I've always believed that the primary responsibility of public officials is to protect people's well-being. Every day that politicians spend distorting science to protect the interests of fossil fuel companies, people die. And unless we act, as populations increase, the death toll from fossil fuel pollution will continue rising.

Fighting climate change by speeding the transition to clean energy is one of the most important things we can do to improve public health around the world. This connection between public health and the planet's health is what drew me to become involved in environmental issues. And it's the most important reason why we must act without delay to fight the sources of both air pollution and carbon pollution. But it is far from the only reason.

## RISING SEAS

Sea levels have risen and fallen throughout history, but until now those changes have happened exceedingly slowly, the only exception being sudden events like the glacial lake burst that Carl described in the last chapter. In 1900, sea levels had been stable for thousands of years. That period of stability is now over. The rising heat in our atmosphere is melting polar ice and glaciers, causing sea levels to rise. Warmer temperatures also cause water to expand, further elevating sea levels. Never in recorded history have the seas risen as much, as quickly, as they did during the twentieth century.

It's easy to be deceived about the potential dangers, because the numbers look small—but they can start adding up fast. Since 1900, global sea levels have risen at a rate of about 0.6 inches per decade. But since 1992, they've risen about twice that fast—and projections for just how much more they will rise vary. It depends largely on how much of the ice in Antarctica and Greenland melts. That, in turn, depends on how much and how quickly we're able to slow our emissions of greenhouse gases. If all the glaciers in the world were to melt, sea levels could rise as much as 230 feet, putting most of the world's population centers underwater. Some politicians like to pretend that the rise isn't already happening, but you don't have to be a scientist to see the proof.

South Florida is one of the places most vulnerable to sea level rise. Most of the Florida peninsula is at or only slightly higher than sea level. To make matters worse, South Florida's cities sit on a landmass composed of porous limestone, which is full of holes. So water not only enters from around South Florida's cities but also from underneath. For these reasons, high tides have always posed a threat to Miami Beach. And as the tides rise, more of the city is threatened.

According to the U.S. Geological Survey, sea levels along parts of the East Coast will rise three to four times faster than the global average, and South Florida is already feeling the effects. A 2016 University of

Miami study found that high-tide flooding in Miami Beach has increased by 400 percent since 2006. Much of this is "sunny day" flooding, meaning that it is not due to a storm but takes place even when the weather is mild. It's becoming increasingly easy to predict when and where Miami is going to experience this kind of flooding, based on the tides and the phases of the moon, and more and more Floridians are being forced to take precautions, like moving their cars to safe places and protecting homes with sandbags. Flooding damages homes, businesses, and cars, backs up sewer systems, and creates traffic jams. Floodwaters also gather sewage and dirt before washing back out to sea, contaminating beaches and bays.

To the city's credit, it is taking action. Miami Beach has earmarked hundreds of millions of dollars for flood mitigation, installing water pumps, raising streets, and building seawalls. But no city has unlimited means, and contending with rising seas will come with rising costs that taxpayers will have to shoulder. In one preview of things to come, in 2014 Miami Beach raised stormwater rates for residents by 84 percent to help pay for the new pumps. But these are merely temporary fixes that won't solve the city's problems long-term. Massive investment will be needed if the flooding continues to worsen—and unless we act, it most certainly will.

United States cities generally measure their vulnerability using federal flood maps. A 100-year floodplain is an area affected by an event so severe that it is likely to occur only once every hundred years. Or: every year, there's a one-in-a-hundred chance that a flood of such magnitude could happen. Although a somewhat arbitrary time frame, it nonetheless provides a good measure of the likelihood of risk.

The flood maps also serve an immediate practical purpose. Created by the Federal Emergency Management Agency (FEMA) to provide a framework for the National Flood Insurance Program, the maps help Americans living in flood-prone areas protect themselves with subsidized insurance, since traditional homeowners' insurance often doesn't cover flooding. FEMA's maps also provide valuable public safety information that city officials can use when storms are approaching—and they offer a good frame for measuring how risks to cities could increase over time in the years ahead. Unfortunately, they can only provide a benchmark, because Congress doesn't allow FEMA to create forward-looking flood insurance maps that take climate change into account. Some members of Congress are almost as afraid of science as they are of losing reelection. Almost.

Fortunately, Congress cannot stop independent scientists from making projections, and these have been unsettling. For example, according to a study by the Boston Harbor Association, a once-in-a-hundred-years storm would inundate about 7 percent of Boston with water, mostly along the edges of the waterfront. By 2100, as a result of rising sea levels, that much of the city could flood *twice daily* during normal high tides. By then, if we don't change course, a bad storm could flood Back Bay and most of the South End and the Financial District, Faneuil Hall, Mass General Hospital, and Fenway Park. In Cambridge, rising seas could flood the campuses of MIT and Harvard. The Boston Museum of Science could be flooded, too.

Compounding the problem is that many of FEMA's current maps are no longer, in fact, current. New York City's 100-year floodplain, for example, had been defined by FEMA in 1983. When our administration took office in 2002, the zone had not been updated since then, despite the fact that the seas had risen and the city had grown. We decided to take matters into our own hands. The New York City Panel on Climate Change, which we had created through PlaNYC in partnership with Columbia University and the Rockefeller Foundation, provided us with the most complete local projections for climate change in any city in the world. With the panel's findings, we created a new flood map for the city—then projected outward, taking into account the most updated information about rising sea levels. Our projections showed that by the 2020s, New York's 100-year floodplain could expand in area by 23 percent. By the 2050s, our 100-year floodplain could include one-quarter of the city. More than 800,000 New Yorkers already live in our projected floodplain for the 2050s—more people than live in all of Boston.

New York, Boston, and Miami are far from alone. Millions of Americans live in homes that could be flooded by normal high tides with a moderate rise in sea level. Two-thirds of the world's population lives in coastal areas, including residents of some of the world's largest cities. Mumbai may have the largest concentration of people at risk from sea level rise. Of the city's more than 20 million people, 2.8 million already live in areas that flood, but by 2070 the number may exceed 11 million.

These numbers matter to everyone—even to people around the world who live far from urban areas. Why? Because nations need strong cities in order to thrive. More than 80 percent of global GDP is generated by cities, and that number will continue to grow as more and more of the world's population become city dwellers. Cities provide people with access to jobs, health care, schools, and other critical resources. Although many don't realize this, the growth of cities is a big reason why the world has seen a sharp drop in extreme poverty over the past quarter century. But climate change threatens that progress.

Take Dhaka, Bangladesh, one of the fastest-growing cities in the world. Its population has grown from an estimated 6 million in 1990 to more than 17 million in 2016 and is on pace to reach 23 million by 2025. Its growth, like that of cities around the world, has been fueled by people looking for an opportunity for a better life. But, increasingly, people are coming for another reason: Climate change has forced them from their homes.

When rising seas inundate land, they destroy homes and crops and render land infertile. Seawater infiltrates fresh drinking water supplies, leaving people without the most important resource they need to survive. Seawater also ruins water used for irrigating crops. Roads, power supplies, and other critical infrastructure fail. In communities that depend on agriculture, people lose the means to feed their families.

Each year, hundreds of thousands of people are migrating from coastal and rural areas of Bangladesh to the capital, many because of increased flooding and extreme weather. The majority arrive in the city poor and without jobs, finding refuge in flimsily constructed homes that are often near bodies of water. According to the International Organization for Migration, 70 percent of those in Dhaka's poorest areas were forced there because of environmental disruptions.

Like South Florida, Bangladesh is particularly vulnerable to rising seas because of its topography. Much of the country is at or below sea level and lies on a delta crossed by many rivers. Its position is even more precarious because it lies at the foot of the Himalayas, where glaciers are melting, further swelling rivers and the Bay of Bengal. The melting of Himalayan glaciers poses other serious risks: They are the source of drinking water for hundreds of millions of people in India, Bangladesh, Nepal, and Pakistan. If they vanish due to warming temperatures, it may cause a water crisis the likes of which the world has never seen.

Cities around the world are facing similar challenges. "We are continuously experiencing both floods and droughts, and this has sometimes resulted in the loss of life," says the mayor of Accra, Ghana, Alfred Vanderpuije. "That's why we have made a strong commitment to addressing climate change in our city."

Ultimately, the impact of sea level rise will depend on our ability to both slow it down and adapt, a topic we'll focus on a bit later. In places like Miami Beach, it's a race against time, and the clock is ticking. In other places, like Bangladesh, the hour of reckoning has already arrived.

## SEVERE HEAT

The year 2014 was the hottest in recorded history—until 2015 claimed the title by the largest margin ever. Then, 2016 beat the 2015 record and brought us another ominous sign: the lowest levels of Arctic sea ice ever observed. Each of the sixteen years we've completed so far in the twenty-first century has ranked among the seventeen hottest years on record.

Some ask, what's so bad about a little more heat? In the middle of February in New York City, a hotter planet doesn't sound like such a bad proposition. And in the summer, what's another couple degrees if it's already hot outside? Isn't that what air-conditioning is for? But heat doesn't only increase the chances of extreme weather by supercharging the atmosphere with energy. Heat itself is a serious risk, and one that already takes a deadly toll.

Many of the deadliest heat waves in recorded history have happened since 2000, and together they have killed more than 125,000 people around the world. In fact, for all we hear about violent storms, heat waves are actually the deadliest kind of natural disaster in the United States, killing more people each year on average than hurricanes, lightning, tornadoes, earthquakes, and floods combined—and that number is increasing with the concentration of greenhouse gases in the air. In 2015, India's second-worst heat wave ever claimed around 2,500 lives. By some estimates, Russia's 2010 heat wave was responsible for more than 55,000 deaths, and the record-breaking heat wave that hit Europe in August 2003 killed thousands more. According to the World Meteorological Organization, heat fatalities during the period 2000–2010 were up more than 2,000 percent compared to the previous decade. These deaths are also probably underreported, since heat is a contributor to many different causes of death.

If we don't hit the brakes on climate change, the number of extremely hot days will climb, and so will the number of deadly heat waves. We're already seeing this happen. In 1950, there was about the same ratio of record high temperatures and record low temperatures recorded each year around the world. Over the last decade, record highs happened twice as often as record lows.

And not everyone has the option of turning up the air-conditioning.

Hundreds of millions of people earn their livelihood working outside, and billions more depend on the goods and services provided through that labor—most critically, food. Above a certain combination of heat and humidity, it is physically impossible to work outside, because the human body can't shed excess heat through evaporating sweat, and that failure can result in heat exhaustion, dehydration, strokes, and even death. We can expect to reach this threshold more and more frequently—not just in the tropics, but in many parts of the United States.

Already a serious public health issue, heat will also become a major economic risk. Even at the lower end of extreme heat conditions, people grow tired more easily, dehydrate faster, and work more slowly, which drives productivity down. Around the world, the number of working hours lost to excessive heat has been steadily increasing. "It's going to get a lot hotter in the United States over the next one hundred years, and worse going forward," says Al Sommer, former dean of the Johns Hopkins Bloomberg School of Public Health. "Montana's summers will soon be the same as New Mexico's today."

Those most affected by the heat will be the world's poor. Many people will have to choose between putting their safety at risk and putting food on the table. And some of the countries that can least afford it could lose billions of dollars in GDP because of shrinking productivity.

Meanwhile, just as rising sea levels are forcing people from their homes, rising temperatures could render parts of the world unlivable. Since 1970, the average annual number of extremely hot days has doubled in North Africa and the Middle East. By the end of the twenty-first century, that number could multiply fivefold. The number of heat waves could increase tenfold. The hottest days could routinely reach 50 degrees Celsius—about 122 degrees Fahrenheit. The Middle East and Africa are home to more than half a billion people. Like rising sea levels, rising global temperatures could trigger mass migrations from the countryside into cities, which could lead to a host of pressures that have the potential to boil over into violent conflict.

## POLITICAL INSTABILITY

In 2010, the UN warned that the four-year drought affecting Syria—one of the worst in the region's history—was causing major crop failures, threatening food supplies, raising food prices, and forcing many into extreme poverty. Hundreds of thousands of people, many of them subsistence farmers, pulled up roots from rural areas and headed to cities, including the capital city of Damascus.

Discontent with the Syrian government had long been simmering, and the effect of the drought was like throwing gasoline on a fire. Sudden upticks in poverty, unemployment, and overcrowding in cities exacerbated preexisting tensions and helped fuel the uprising in 2011, which has since tragically devolved into the worst humanitarian crisis of our time.

Climate change didn't cause the Syrian crisis, which is the product of many complex factors. But the devastating effects of the drought—likely worsened by climate change—helped spark unrest. Riots in Tunisia and Egypt that precipitated revolutions in both countries were fueled in part by increases in food prices. That region of the world is not alone in facing these risks. Climate change could affect food security and economic security not for millions but for billions of people.

Consider the crop that launched the first agricultural revolution: wheat. Wheat is the single most important crop to the global food supply. Research now projects that every 1 degree Celsius increase in temperature will cut global wheat yields by roughly 7 percent. Even a small decrease in wheat yields could mean higher food costs and lower revenues for farmers, which could lead to more poverty and hunger, especially as the global population grows.

Many people depend on farming not only as their source of income but as their primary source of food—and nowhere will climate change have a bigger impact on these farmers than in sub-Saharan Africa. Africa is more economically dependent on agriculture than many other parts of the world, and most of sub-Saharan Africa's agriculture is subsistence farming and heavily dependent on rainfall. Only about 7 percent of the continent's farmland is irrigated, making Africa's farms, and the people that depend on them, especially vulnerable to changes in weather. Over the last decade, the region has made great strides in raising living standards and fighting poverty. Climate change threatens to reverse those gains and leave people without a way to feed their families.

The slight shifts in weather that can harm subsistence farmers also threaten big commercial growers. Billions of dollars of capital invested over the last century in developing global supply chains of crops like cocoa and coffee are now at risk from climate change. In fact, coffee production in many regions is already slumping under the impact of hotter temperatures and other effects of climate change. Higher prices combined with job losses and shrinking incomes could be a recipe for political disaster in many parts of the world.

Environmental changes can upend societies. That's why the U.S. Department of Defense, hardly a bunch of tree-hugging environmental activists, has called climate change a "threat multiplier" that weakens "the ability of governments to meet the basic needs of their populations." The department's job is to keep our country safe by identifying and protecting us from emerging threats. That requires planning for all potential contingencies and unexpected developments, both in the near and the long term. How could it not consider climate change?

# OCEAN LIFE

As we pump $CO_2$ into the atmosphere, it doesn't all stay there. Much of it is absorbed by oceans. Since the dawn of the industrial era about 250 years ago, the oceans have absorbed about half of all $CO_2$ emissions.

Oceans sucking carbon out of the air may sound like good news. It isn't. When $CO_2$ is absorbed by the oceans, it dissolves and becomes carbonic acid. As the concentration of $CO_2$ in the atmosphere increases, the oceans absorb more of it and become increasingly acidic. It's estimated that $CO_2$ produced by humans has increased the acidity of the oceans by nearly 30 percent. Higher acidity triggers chain reactions that ripple throughout marine ecosystems. When $CO_2$ reacts with water, it robs the ocean of compounds that many species use to build their shells and skeletons. Acidic water causes shells to dissolve. It also damages corals that nourish the ocean and provide habitat for millions of species.

This is especially troubling because today around three billion people rely on fish as their primary source of protein, or as a source of income. By 2030, the global demand for fish is expected to increase by about 20 percent or more. At the same time, however, the number of fish in the ocean is being depleted because of destructive fishing practices and inadequate policies, and climate change might threaten the very survival of many species.

While global temperatures are rising, more than 90 percent of the increased heat we've created over the last century has ended up being stored in the oceans. This is a reason why sea levels are rising, since warmth causes water, or almost any substance, to expand. Higher ocean temperatures also increase the intensity of cyclones, which damage marine ecosystems, fisheries, and coral reefs. Rising temperatures are causing migrations on land, but these changes are happening even faster in the seas. Already entire populations of certain species have migrated from areas they have long occupied because of changes in ocean temperatures.

One of the most dramatic and visible symptoms of rising ocean temperatures is what's called coral bleaching. Coral reefs get their normal healthy color from resident algae that are the coral's main source of food. When water temperatures get too high, coral expel the algae, leaving themselves vulnerable to disease and starvation. Australia's Great Barrier Reef, the largest coral reef in the world and one of its most spectacular and biologically diverse places, is undergoing bleaching so severe that scientists have warned that the reef is nearing "complete ecosystem collapse." It's hard not to see that this is a public health and economic issue as well as an environmental one. If supplies of fish continue to fall, millions of people will lose their livelihoods. Many will face hunger—or starvation.

In recent years, Bloomberg Philanthropies has been working to protect fish as a food resource, mostly by reversing the damage done by overfishing and destructive fishing practices. We've been collaborating with three countries that are home to 8 percent of the world's fisheries by volume: Brazil, Chile, and the Philippines. Smart practices are proven to help fish populations rebound dramatically and strengthen local economies at the same time. Often this comes down to good management—like leaving certain areas unfished for a certain period of time, much as farmers leave fields fallow to regenerate—and better enforcement by authorities. But all of these efforts could be rendered moot if we continue emitting greenhouse gases into the atmosphere at the rates we are today.

It's not just ocean ecosystems that are being thrown out of equilibrium by climate change. Warmer temperatures set off changes that ripple through the food chain. And as cooler regions become warmer, they also become susceptible to infectious diseases. Hotter temperatures will expand the areas in which mosquito-transmitted diseases like Zika, West Nile virus, and dengue can thrive.

All of the changes described in this chapter are just the tip of the iceberg (to use an endangered metaphor). Just as no weather event exists in a vacuum, all of the risks we face from climate change are intertwined. They complicate and compound one another, and no one knows exactly where this road will lead.

It is true that the climate has undergone many dramatic changes over the Earth's history. But it is also true, as Carl has explained, that our current global civilization was made possible because of a period of unusual stability, and never before have human settlements been so large or so permanent. Nomadic communities of hunter-gatherers can move when their way of life is threatened. But we can't pick up and move Miami, or New York, or Guangzhou, or Mumbai. No one can afford to ignore this reality, because while the effects of climate change are felt locally, their repercussions are global. Altering the stability of the climate could carry profound—and unpredictable—political, ecological, economic, and humanitarian consequences.

All the data indicates that we're already living in times of more turbulent climate—and that if we do nothing, there's far worse to come. We have the opportunity to avoid these harmful effects, while also immediately improving the lives of millions of people. We'd be crazy not to embrace it.

# PART III

# COAL TO CLEAN ENERGY

5

# COAL'S TOLL

*Our families deserve clean air, and we have been without
it for far too long.*
—Michigan resident Alisha Winters, writing
to her local utility about a coal plant
in River Rouge, MI

On July 21, 2011, I was standing on a ferry in the middle of the Potomac River, near Alexandria, Virginia. It was midday, sun blazing, oppressively humid, and hot as hell—99 degrees. Whose idea was this, anyway?

With our backs to one of the country's worst-polluting power plants, the GenOn Potomac River Generating Station, I committed $50 million to closing or phasing out one-third of the nation's coal-fired power production by 2020. The Sierra Club's Beyond Coal campaign to push the U.S. economy toward cleaner energy sources had already had some success, on a small scale. I had never before given a major gift to an environmental group, but Carl had convinced me to back a huge expansion of the campaign. I'm not one to tilt at windmills; I like to fight battles that are winnable. In this case, however, I was fighting *for* the windmills (and the solar panels). It was an ambitious crusade well worth supporting.

I'm an unrepentant capitalist, not exactly the Sierra Club's target demographic. But I also believe that government's most important duty is to protect public health and safety. When the push for profits endangers public health, I don't have much sympathy for industries whose products leave behind a trail of diseased and dead bodies. That doesn't mean I want to outlaw all of them. I've stood up for the right of tobacco companies to exist, for instance. But for everyone's sake, we should aim to put them out of business by driving down demand, through taxes that address their true societal cost, regulations that mitigate the harm they cause, and public awareness advertising that makes plain the dangers they pose. The same is true of coal companies.

In the chapters ahead, we'll examine how to address the major sources of climate pollution that are warming the planet. It makes sense to start with the biggest: coal.

There was a time when we needed coal. From the days of Thomas Edison until fairly recently, coal was the primary source of U.S. electric power, providing up to half of our nation's electric supply (a mix of other sources, like petroleum products, natural gas, hydroelectric, and, later, nuclear and other renewables, provided the rest). Cheap, accessible, and plentiful, coal brought light into homes and powered railroads, steam engines, and factories. It seemed like a godsend—and, for well over a century, a coal-based energy economy seemed inescapable. After all, what was the alternative?

Coal's role in powering the nation's rise to greatness also gave it a special place in the American imagination. The image of rugged, blackened coal miners in Appalachia was a symbol of our growing industrial power. But romanticism has a way of concealing what we don't want to see, and coal's dark side has always been ugly: mine cave-ins and explosions; child workers and violent labor riots; black lungs and early deaths; soot-filled air and toxic water.

In the United States, coal mining accidents regularly killed at least one thousand miners annually through the 1940s. Globally, the numbers remain high. In China, an average of 20 coal miners died each day from 1996 to 2000, an annual toll that exceeded 7,000. It was seen as cause for celebration when the Chinese government reported that coal mining deaths had dropped below 1,000 (to 931) in 2014.

"A canary in a coal mine" is a phrase we now use to refer to anything that serves as a preemptive warning of danger. But it refers to the practice of using canaries to see if a mine's methane or carbon monoxide levels had become too dangerous for work. If the canary stopped singing, miners knew they had to evacuate the mine or stay out of it. And while industry safety practices have improved, the costs that coal imposes on society have remained tragically high.

I knew coal was a dirty fuel, but I hadn't realized just how deadly it was until Carl showed me the numbers in 2011. They were staggering. Coal pollution was prematurely killing 13,200 Americans a year—36 people *a day*—through respiratory disease, lung cancer, and other illnesses. Another 20,000 Americans had heart attacks related to coal pollution every year, and 217,000 had asthma attacks. The financial toll in annual health costs exceeded $100 billion. In Europe, coal power production causes over 22,000 premature deaths a year. In India, it causes 100,000 premature deaths a year. Every new coal-fired power plant in Indonesia is projected to kill more than 24,000 over its forty-year lifetime.

Burning coal is the largest source by far of the toxic mercury contaminating our fisheries, making it dangerous for pregnant women to eat fish from many bodies of water. And 2,000 miles of streams, creeks, and rivers in Appalachia are already gone, or will no longer exist, because they have been filled with waste from mountaintop removal mining.

Burning coal leaves behind fly ash, which for many decades was simply released into the air. Now, federal regulations require it to be captured, but the captured ash is stored in pools that can be just as damaging to the environment—especially if the liquefied coal ash in the pool, called "coal ash slurry," escapes. Imagine an oil spill, but with only the most concentrated and polluted oil. Now imagine it on land. That's what a slurry spill is like. It destroys natural habitats, seeps into ground-water, and flows into rivers. The EPA estimates more than 23,000 miles of rivers and streams have been polluted in this way. Half the toxic chemicals polluting America's streams and rivers come from coal ash pools.

As bad as coal is for our health, it's also the biggest single greenhouse pollutant. In the United States, it accounts for almost a quarter of our greenhouse gas emissions.

## JOINING THE FIGHT

I first learned about the Sierra Club's Beyond Coal campaign by acci-dent. At a meeting one morning, I listened to a proposal on education reform from a group seeking a major gift. The idea didn't seem to click, so I asked my staff if there were any other areas we should be looking at. As it happened, one of my deputy mayors, Kevin Sheekey, had just had lunch with Carl, who had begun fund-raising to expand Beyond Coal. Kevin was impressed with what he heard. So we set up a meeting.

Carl laid out the public health case against coal, which was over-whelming. He then explained the political case, which was no less con-vincing: With cap-and-trade legislation all but dead in Congress, closing coal plants was the most practical way to make a difference on climate change. Next came the economic case, which also made sense: Even though the cost of renewable energy was falling, there was a serious dan-ger that the coal industry could expand its scope for decades if utilities made the mistake of building new coal plants and signing new long-term contracts with aging ones. Those contracts would have locked in decades of public health problems and might have permanently tipped the scales in the effort to fight climate change.

The economic, political, and public health arguments in favor of action were all clear, but then Carl got to the actual strategy—and it was pretty thin. I've always liked David versus Goliath fights, and I've picked (and won) more than my share of them. But before I engage, I need to be able to see a path to success—and establish the benchmarks by which we'll measure it. The Sierra Club was used to operating on passion. But if passion isn't carefully directed and managed, a lot of energy and money gets wasted, and a lot of frustration sets in.

While extremely spirited, the Beyond Coal team was not particularly methodical about its work. They had no quantitative analysis for which plants they targeted for closure, and no metrics for measuring progress. Our staff at Bloomberg Philanthropies pushed the Sierra Club to come up with a strategic plan and spent several months evaluating its campaign. The Sierra Club team did a plant-by-plant analysis, assigning metrics for environmental impact and political climate (that is, how much opposition we anticipated) to determine the expected value of targeting any given plant. They then organized plants into three categories: easy, moderate, and hard. Our foundation staff also conducted its own analysis, which lined up with the Sierra Club's. Together, we asked ourselves: "If we think we can win this many easy cases, this many moderate ones, and this many hard ones, what is our outcome? What is a realistic timetable? What will it cost? And will the results be worth it?"

After answering those questions to our satisfaction, Bloomberg Philanthropies made the $50 million grant to the Sierra Club, which allowed the campaign to expand from 15 states to 45. The funding also enabled the club to begin developing better data and analytics to help us take on the coal industry more methodically. Our goal was to retire one-third of the U.S. coal fleet (in other words, to cut coal plant capacity by one-third) by 2020.

In one early battle, we took on two old, outdated coal plants that had been sickening neighborhoods in Chicago. We estimated that they were causing 720 asthma attacks and 42 premature deaths a year—plus $120 million in health costs. What was worse, a number of park sites used by local children, including a swimming pool, baseball field, and playground, sat directly underneath one of the plant's smokestacks—"the cloud maker," as some of them called it. The owners agreed to a closure announcement that would have let them operate for another ten years. Not good enough, we said, and, working with Mayor Rahm Emanuel, we brokered a deal to close both plants in 2014.

In that case we had a crucial ally in Mayor Emanuel. But in most other cases, a public health argument alone isn't enough. One of the biggest reasons Beyond Coal has been so successful is that our citizen-activists don't just tug on heartstrings; they also focus on purse strings. In most cases, the economics are not on coal's side.

The average U.S. coal plant is more than fifty years old, which is well past the originally anticipated life expectancy. Upgrading pollution controls and installing scrubbers (which remove particles from exhaust) is almost always more expensive than simply replacing plants with clean energy options. Yet utilities often prefer to keep old plants open, sticking ratepayers with the bill. Beyond Coal helps communities to argue that the smartest financial path forward for utilities increasingly leads away from coal and toward solar, wind, or natural gas. Economics can be an environmental advocate's best friend.

In Kentucky, for example, American Electric Power (AEP) had proposed a $1 billion retrofit to bring one of its coal plants up to EPA pollution standards. To pay for it, the company wanted to pass the costs on to ratepayers by raising electric bills by around 30 percent. But the alternative—converting the plant to natural gas—was both cheaper and cleaner. Facing opposition from local business owners and consumers, as well as the state's attorney general's office, AEP eventually closed part of the plant and converted the rest to natural gas.

# THE (MARKET'S) WAR ON COAL

Politicians use the term "war on coal" as though coal companies were the victims of an unjust attack. Unfortunately, environmentalists took the bait and have been playing defense. Well, let's get it straight: It *is* a war—a war to protect innocent lives from unnecessary disease and early death, and the environment from severe harm. Not surprisingly, those who criticize the war on coal never mention the actual death toll. And the trouble is, too often, environmentalists don't, either. Climate-change campaigners often speak in incomprehensible technical terms, rattling off numbers—tons of carbon, parts per million—that are completely meaningless to most people. It is far more effective to be able to point to the worst power plant in America and say that it kills 278 people a year and causes 445 heart attacks. When you explain that to the public, they understand that this is a war worth fighting—and one we must win.

Coal's supporters counter by saying that coal energy is cheap energy, and cheap energy fuels the American economy and our prosperity. That argument was long true, but not anymore. In 2010, coal power was cheaper than wind and a lot cheaper than solar. Today, coal is simply not competitive. I'm very sympathetic to those who work in the coal industry, and governments ought to do more to help them find other career opportunities. But the fact is: governments and nonprofit organizations are not leading the war on coal. The market is. And when the market wages war, it takes no prisoners.

There are four main reasons why coal is no longer economical: the boom in natural gas production, advances in technology, improvements in pollution-reduction regulations, and shifts in consumer preferences.

**New Gas.** Natural gas, when safely and responsibly extracted, has been a godsend for the environment and public health. A decade ago, natural gas power plants in the United States sat idle because gas was too expensive. Energy companies were building terminals to *import* liquefied natural gas from overseas. The shale revolution changed that. There has always been a lot of gas-rich shale rock in the United States, but conventional wisdom was that shale gas was too expensive to extract, and that the country was therefore running out of available domestic gas. But starting in the early 2000s, natural gas companies had a major breakthrough: They figured out how to efficiently separate gas from dense shale rock through hydraulic fracturing, or "fracking."

This development turned natural gas from a scarce, expensive commodity into one plentiful enough to make America a net gas exporter. In a single year, 2008–9, the price of gas fell in half, from $8 to less than $4 per unit. Those idle gas power plants became useful again, and by 2016, for the first time, the United States produced more electricity from natural gas than it did from coal.

The spread of gas drilling has spawned an anti-fracking movement among many environmentalists. Fracking, like any extraction technique, requires safeguards, and drilling shouldn't be allowed everywhere, including sensitive areas like watersheds. But fracking allows for the most efficient extraction of natural gas, and as long as we need natural gas, it makes sense to frack.

While gas burns cleaner than coal, unburned natural gas is an even more potent source of greenhouse gas emissions. If we're going to realize the potential of natural gas, methane leaks must be reduced, if not eliminated. Using the best practices and done correctly, most of the negative effects of gas extraction can be dramatically minimized. But gas companies haven't—and on their own won't—uniformly adopt these best practices.

That's why Bloomberg Philanthropies teamed up with the Environmental Defense Fund in 2012. We both saw that fracking needed a better regulatory framework, so we worked together to develop one that would protect local communities and the environment by reducing methane leaks, groundwater pollution, and the minor earthquakes that sometimes result. That doesn't mean common sense regulations aren't also in the best interests of the energy companies. After all, companies that leak methane also leak profits. And when we worked in states like Colorado *with* energy companies, we found we had a lot of common goals.

Critics contend that natural gas is still a fossil fuel and is only prolonging our reliance on carbon. They're right that gas is not a permanent solution. Even if we replaced all the coal we burn with natural gas, we would still overshoot the atmosphere's expected tolerance for CO2, only more slowly—and we can't allow that to happen. But what the critics forget is that we can't simply turn off the lights—which is what would happen were we to shut off all nonrenewable energy sources. We just don't have the renewable capacity yet. Until we do, natural gas is an essential alternative.

We should keep working to make natural gas cleaner and fracking as safe and clean as possible. Until renewable capacity can meet demand, however, natural gas is not only the bridge fuel we need to reach a decarbonized economy, but a major force in reducing the coal industry's market share. If we're going to win the war on coal, we need to have natural gas in our arsenal.

**New Technology.** The cost of clean energy has plummeted in price in recent years. The U.S. Department of Energy reports that from 2008 to 2016, the costs dropped 41 percent for wind, 64 percent for solar, and as much as 94 percent for LED lights. Two-thirds of new electricity capacity installed in 2015 was wind or solar. The drop in price follows a trend that occurs as any new technology improves and advances. In the 1960s, the first electronic calculator cost thousands of dollars and weighed about fifty pounds. The first pocket calculators, introduced in the early 1970s, cost several hundred dollars. By 1978, they were selling for ten dollars. Now they are free to anyone with a Wi-Fi connection. Solar panels, wind turbines, batteries, and LEDs are on a similarly sloped downward price curve. As that happens, public and private investment in renewable energy is growing.

Government policy has also played an important role in spurring technological advancement in renewable energy, just as it has in other industries. The federal government provides research and development funding for solar and wind technology, and it gives wind and solar projects tax benefits (though not as generous as the benefits awarded to the fossil fuel industry). States also established targets for renewable energy, which led utilities to build clean technology projects. Other countries played critical roles, too. Wind turbines first got to scale in Denmark, and it was the Germans and then the Chinese who brought down the cost of solar panels through subsidies that allowed the young industry to grow.

As solar and wind technology have become more cost-competitive compared to coal, investment has flowed from the latter to the former—and jobs have, too. There are now far more people employed in the renewable energy industry than in the coal industry. "Clean energy jobs are some of the fastest-growing in Tennessee, nearly triple the state's overall employment growth," says Madeline Rogero, the mayor of Knoxville, Tennessee. "These are the energy jobs of the future."

In 2015, clean energy investment outpaced oil and gas investment for the first time. In response, several energy companies have filed legal motions to lower the appraised value of their coal plants in an effort to lower their tax burdens. Their argument: in light of the glut of natural gas and the competitive prices for solar and wind power, coal plants just aren't worth what they used to be.

In fact, some of the nation's newest, ostensibly cleanest coal plants are white elephants: After spending $750 million to modernize the Homer City Generation Station, one of Pennsylvania's largest, GE wrote off $800 million of its total $2 billion investment just a year after reopening it. Three years after Colorado's utilities sorted out which of their coal plants were worth cleaning up, seven of the ten they chose to retain already cost more to operate than it would have taken to replace them with new wind farms. While successfully arguing to devalue a three-year-old coal plant in Texas to $432 million from its original $1.7 billion valuation, an attorney for the plant told the courtroom that it would be the "last coal plant in Texas." Let's hope so. He summed it up nicely when he said, "Clean energy and natural gas are the future."

For that future to become reality, sometimes we just need government to get out of the way. For instance, the Obama administration, hoping to protect the nascent U.S. solar industry, imposed tariffs that could exceed 200 percent on solar panels from China and Taiwan. It was a gift to a special interest that hurt both our economy and our ability to fight climate change. Since installing solar panels produces more jobs than making them, restricting the supply of solar panels in the United States was counterproductive. In all likelihood, the tariff led to job losses. In addition, it made solar panels more expensive, slowing the uptake of renewable energy. Cheap Chinese panels *are the main reason* the cost of solar has plummeted. We need to take advantage of those types of innovations, whether they come from China, Germany, or Ohio. Erecting protectionist trade barriers will only prolong the transition to a low-carbon economy.

**Proper Regulation**. For over a century, mining and energy companies have been able to privatize coal's profits while socializing its costs. Translation: corporations make money and taxpayers pay the price. This is still largely true. And yet the many ways in which mining and burning coal impose costs on the rest of us are rarely recognized. Coal companies pay neither to care for the people their plants sicken nor to clean the air of the toxins their plants emit, to say nothing of the huge climate costs imposed by carbon emissions. In a properly functioning market, coal companies would have to account for the costs they impose on society. Instead, those costs are still largely borne by society.

Coal's ability to make the rest of us pay its bills didn't come about accidentally. All the way back to their early days, coal companies have largely written their own rules and blocked efforts to clean up the industry. When Congress wrote and amended the Clean Air Act in the

1970s, environmentalists argued that existing coal-burning power plants should be required to install modern pollution-control equipment. Coal lobbyists blocked the move, promising that since old coal plants would soon be replaced with modern ones, there was no need to regulate them. Unsurprisingly, many of those plants operated for decades to come, and some are still in business.

Government did little to clean up power plants over the next four decades. After failing to pass a cap-and-trade bill in 2010, the Obama administration adopted the Clean Power Plan in 2015, requiring states to reduce emissions by an average of 32 percent below 2005 levels. That would be important progress, but it sounds more impressive than it is. Why? Well, in 2015, emissions were *already* 21 percent below 2005 emissions, thanks mostly to the retirement of so many coal plants. So the real goal of the Clean Power Plan is only an 11 percent reduction from 2015 levels by 2030. In other words, the plan's goal is to cut emissions at less than half the rate we achieved over the past decade. Not exactly radical. In fact, we would likely hit the goal even without the Clean Power Plan, as coal plants continue to close.

Still, the Clean Power Plan serves as a useful prod to states and utilities that would help strengthen the market forces already driving coal-fired power plants out of existence. It is just less ambitious than it should have been. And while the Trump administration may gut it, it's worth remembering what happened from 2010 to 2015, when emissions fell and so many coal plants closed: The price of wholesale electricity fell by a quarter, and there was no increase in the frequency of power outages. We can cut costs *and* emissions without sacrificing reliability—or human lives.

**Consumer Demand.** The market operates on the basis of consumer demand, and the fact is that when given the choice, consumers, both Democrats and Republicans, would rather buy clean energy than dirty energy. The reason is simple: They do not want the air they breathe and the water they drink to be polluted with toxic chemicals. That is why citizens all over the country have joined together in pushing their local utilities to phase out coal plants and replace them with cleaner forms of energy.

Over the past six years, the Beyond Coal campaign has led to the closure or phase-out of more than 240 plants—one-third of the total U.S. capacity. Individuals, as consumers who are concerned about their pocketbooks and as citizens who are concerned about the health of their communities, have been a driving force in coal's rapid decline—and a major reason why the share of U.S. electricity generated by coal has dropped from half in 2005 to less than a third in 2017.

A good example of how Beyond Coal has worked comes from Omaha, Nebraska. The Omaha Public Power District (OPPD), a local utility, had owned and operated a large coal plant, its North Omaha Station, for more than sixty years. Since it had been built well before the Clean Air Act, it was not equipped with technology to limit its pollution. The act had stipulations for this, of course, mandating that the plant update pollution controls whenever it modernized the plant. But OPPD never upgraded the plant, allowing it to become one of Nebraska's largest sources of air pollution. A Nebraska state legislative study linked it to 240 asthma attacks, 22 heart attacks, and 14 deaths a year, imposing $100 million in health and environmental costs on the community.

Nebraska's Sierra Club chapter decided that this plant was a prime candidate for the Beyond Coal campaign. The chapter had spent years talking with the community about the plant's health hazards. But it was only when the club demonstrated to the utility that moving from coal would be cost-effective that the situation changed. In June 2014, OPPD approved a plan to retire three of five coal-fired units at the plant by the end of 2016. Meanwhile, they promised to convert the two remaining units to cleaner-burning gas by 2023. Even better: the utility committed to implementing energy-efficiency measures and investing in substantial new wind power. By 2018, one-third of the area's power will come from renewable sources.

## SUBSIDIZING THE PAST

Despite its health dangers and bad economics, government still subsidizes coal in ways that give it a price advantage over renewable and other cleaner-burning fuels. Consider the Powder River Basin, which straddles Wyoming and Montana. The basin contains two of the world's largest coal mines, which alone produce a fifth of the nation's coal. The entire region produces about 43 percent of America's coal, or nearly 500 million tons each year, virtually all publicly owned. Yet, amazingly, the U.S. Department of the Interior has not classified Powder River as a coal-producing region.

If it did, it would have to sell the coal from public lands in a fair market, with competitive auctions. Instead, current rules allow coal companies to buy publicly owned coal at whatever price they set. This not only deprives taxpayers of the true financial value of public coal, but it also enables mining companies to flood the energy market with subsidized coal, unfairly competing with gas and renewables. Taxpayers get cheated out of nearly $3 billion a year in the deal. There was a lot of talk in 2016 about a "rigged" economy. Well, the energy market is rigged in favor of coal. And while coal is still losing badly, it is hurting the rest of us on its way down.

The Obama administration finally began to limit these giveaways in 2016, by imposing a moratorium on new leases of coal on public lands. But existing leases, at far below market value, remain intact. One study estimates that just removing the federal subsidy would reduce demand for Powder River coal by up to 29 percent. Ending the sale of coal from public land would raise the price of coal to market levels, forcing it to compete on a more level playing field with cleaner fuels like renewables and gas.

Because of the political power of the coal industry, the federal government props up coal companies in other ways, too—ignoring their use of creative accounting to shift liabilities (like worker pensions and environmental cleanups) to shell companies they allow to go bankrupt.

By allowing this scam, our legislatures and regulators hurt coal workers and impose costs on taxpayers. Time and time again, as mining companies run out of money, courts have let them take the money set aside for pensions and use it for their own operations, leaving the public to pick up the uncovered costs. In 2015, for instance, a bankruptcy court allowed a mining company to use funds set aside for the pensions of 208 Indiana miners, spouses, and widows to pay its lawyers instead. Yet even after they go bankrupt and abandon their employees and neighbors, coal companies manage to keep paying the politicians who have protected them. In 2016, bankrupt coal companies gave federal and state candidates nearly $1 million from their corporate political action committees. So civic-minded!

## COAL IN THE WORLD

In early 2014, when I accepted UN Secretary-General Ban Ki-moon's invitation to take on a diplomatic role as his Special Envoy for Cities and Climate Change, neither India nor China had indicated a willingness to make any commitments toward a global accord on climate change. Along with the United States, these two countries burn more than 70 percent of the world's coal.

That September, I met with Indian Prime Minister Narendra Modi and offered my support for his effort to develop one hundred "smart cities" in India. A few months later, I flew to India to participate in the country's first renewable energy summit and to meet with Modi and other government officials. Modi had signaled that India was going to be much more ambitious about its renewable energy goals than it had been under the previous government. The renewable energy summit was an early test of whether his government would follow through.

At the summit, I met with India's energy minister, who said that while the government had increased its 2022 renewables goal by 500 percent, he could raise the goal by three or four times if he could attract more private capital. He told me he would say that from the stage, and he was as good as his word. The business leaders in attendance responded favorably, offering to undertake construction of plants that could generate more renewable energy than existed in the entire world at that time.

Beyond public health and climate change, India has another reason to switch from coal and over to renewables: lack of reliability. India's coal-fired power plants often sit idle because they lack a reliable or affordable coal supply, or because there is not enough water to run the plants. So the government has agreed that, for the next decade, it won't build new coal-fired power plants, a pause that may well lead to a longer-term scale-back of the country's coal ambitions.

Globally most coal is used for electricity—that's the focus of the Beyond Coal campaign—but a lot is also used to smelt steel and fire cement, and in China a tremendous amount is still burned for home heating, as used to be the case in the United States and Europe. In fact, China accounts for half of the world's coal production and consumption. But China, like India, has also made important progress in recent years. As public awareness of the dangers of air pollution mounted, the Chinese government moved quickly. It shuttered old coal plants and created new regulations for all new plants. In some regions, it canceled all new coal plants completely.

Thanks to these new policies, some experts estimate that over 110,000 fewer Chinese people will die of pollution-related deaths in 2017, with an additional 2.1 million rescued from debilitating lung disease. The world is better off, too. In no small part because of China's moves, global greenhouse gas emissions have stabilized over the past three years.

China and India both still have a long way to go to wean themselves from coal, but both are making progress, and other countries are moving even more rapidly. The UK has announced that it will eliminate its last coal-fired power plant by 2025. In 2016, there were several brief points at which absolutely no electricity was being generated in the UK by coal—a first since 1882. Portugal is routinely powered by renewables alone for days. And in countries from Myanmar to Chile, citizens are pushing back against coal—and winning. The International Energy Agency found that renewables like wind and solar represented more than half of the total global growth in electrical generation capacity in 2015, and that this trend would accelerate over the next five years.

We still have much work to do at home and around the globe, and not all the news is good. As new coal plants become less marketable in China and India, coal-producing Indonesia is now burning coal at a faster rate, consuming the coal it can't otherwise export. Turkey, too, remains committed to coal power, even though numerous analyses have shown that a wind- and solar-based strategy to reduce reliance on Russian gas would cost far less. Only about 20 countries account for 90 percent of existing and planned coal power. More attention needs to be paid to them, and more support needs to be provided to them, in order to help them make the same transition that is now under way in the United States, where there is no longer any doubt: We are winning the war against coal.

Critics of Beyond Coal and the Clean Power Plan have described them in apocalyptic terms. But much of what the critics proclaimed about the plan and our campaign—that they will destroy the coal industry, kill jobs, and raise costs for consumers—is wrong. Coal is declining under the weight of its own flaws: It pollutes, it's no longer competitive, and, most importantly, it kills people. Meanwhile, market forces, technological advances, and public demand for clean air and action on climate have combined to make alternative sources of energy more financially attractive.

"Coal is the single greatest threat to civilization and all life on our planet," the scientist and early climate change pioneer James Hansen has said. America helped lead the world into the Industrial Revolution. Today, it is helping lead the world out of one of its worst legacies.

In 2015, the Sierra Club and Bloomberg Philanthropies increased our goal for Beyond Coal. We had originally aimed to cut coal power by one-third. Now, we are aiming to cut it in half by the end of 2017. The GenOn Potomac plant—the backdrop for our kickoff announcement on that hot July day—is now closed. And a coal-free world is finally in our sights.

6

# GREEN POWER

*Our municipal utility will move to 100 percent
renewables. . . . Environmental zealots have not taken
over our city council. Our move to wind and solar is
chiefly a business decision.*
—Dale Ross, Republican mayor of Georgetown,
Texas

September 2, 2009, Mumbai. India's advertising industry and the press barons supported by it have gathered in a five-star hotel to launch India's—and I suspect the world's—biggest-ever public-service advertising campaign, titled: Lighting a Billion Lives. The initiative is led by Dr. Rajendra Pachauri, India's leading climate scientist. (Pachauri headed the Intergovernmental Panel on Climate Change, which shared the 2007 Nobel Peace Prize with Al Gore.) That evening I learned something that shocked me: India still had 400 million people with no electricity—and therefore no light at night except from candles or kerosene. Globally, about 1.2 billion people live without electricity, and one-third of them are in this one country.

Part of my shock stemmed from the fact that, almost fifty years earlier, when I'd spent two years living in a poor and remote village in eastern India, I had had light and power—as did my Indian neighbors. An American aid project called the Damodar Valley Corporation, modeled on the Tennessee Valley Authority, had electrified the area. Astoundingly, the light hadn't spread much farther since then. So what had happened in the intervening half century?

The government had envisioned the steps to electrification as follows: First, build big power plants (mostly powered by coal or hydro), then string copper wire to villages, and then run that electricity through incandescent bulbs. This was, it turned out, too wasteful. Unfortunately, only a trivial fraction of the energy in the coal turned into light, and much of the copper wire was stolen (along with a good chunk of the electricity). As a result, a third of the country still lacks any access to modern energy.

Now, there's a better way. Solar panels, small batteries, and LED lighting make it possible to do away with the whole grid and instead light households, minimally, for a one-time cost that seems to average about $200 a family. (A basic solar lantern, which can replace kerosene lamps, costs only $40.) Pachauri's Billion Lives vision is to establish village centers to recharge solar lanterns, and to try to get philanthropy to fund the effort village by village.

Farooq Abdullah, the leading Congress party figure in disputed Kashmir, was the newly minted minister for New and Renewable Energy. He told those of us in the ballroom in Mumbai that the government of India subsidized 2 billion liters of kerosene each year, at an annual cost of about $1 billion.

Scribbling on the back of my napkin, I realized something startling. The Indian government was spending enough each year on kerosene subsidies to replace 25 million kerosene lanterns with solar ones. Every three years it spent enough to give solar power to all of the 75 million Indian households without electricity. In other words, the government was paying far more to subsidize kerosene—which is dangerous and toxic—than it would cost to provide villagers with solar power.

The problem of energy access is never a question of cost or price—something else is broken.

So I began my exploration into what some call Utility 2.0—the electrical system of the twenty-first century. From the time of the Mumbai conference onward, the cost of solar lighting continued to plummet. Panels, LED lights, and batteries got ever cheaper. The price spread solar enjoyed over kerosene or diesel fattened. Yet even with this trend the International Energy Agency concluded in 2010 that, on a business-as-usual trajectory, the number of people without access to light would increase, not decrease, because population growth would outstrip new grid connections. Today, nine years after I learned that more than a billion people lacked electric light, most of them still do—even though the raw economics of giving them solar power have gotten stronger and stronger.

In 2011 Ban Ki-moon, then the secretary-general of the UN, who himself had to study by light from kerosene lanterns when he was growing up in Korea, made providing electricity to all the centerpiece of his global energy initiative. It was the UN's first foray into energy, and since generating electricity accounts for 25 percent of total greenhouse gases, it's a critical piece of the climate puzzle. Yet while innovation is happening rapidly, there are still critical barriers to overcome.

Universal electrification, particularly for rural areas or the poor, has always faced a significant challenge. Generating and distributing power is capital-intensive—the systems cost a lot to build. In the 1930s, 95 percent of American farmers lacked power because their communities couldn't afford these costs. My father, who worked on rural electrification, explained to me over the dinner table when I was a boy that this was why Franklin Roosevelt had decided to fix the problem with a simple solution. The federal government guaranteed that farmers who organized themselves in rural co-ops would be eligible for low-interest loans. Within ten years rural America was electrified.

At the same time, as my father also explained, the Rural Electrification Administration *made* money for the government, because electrifying farms was good business. The price calculations I made that evening in Mumbai on solar versus kerosene suggested that it was even better business in the twenty-first century. Still, solar panels could provide cheap electrons only if credit were made available to overcome the biggest challenge facing poor rural households—up-front costs.

Kerosene is dirty and expensive. But you can buy a glassful—enough for a week of light—in the market. A solar panel is clean, cheap, and provides power for about twenty years, but you have to pay for it all at once. Households with no property title, no bank account, and no credit rating have a hard time borrowing the money to buy a panel. In other words: the poor can afford the daily cost of clean, renewable electricity, because the daily cost over a twenty-year period is minuscule. But they can't finance twenty years worth of energy up front. They need a banker. Theoretically, the multilateral development banks, led by the World Bank and its regional colleagues like the Asian, African, and Inter-American Development Banks, are supposed to be doing that kind of job—helping the poor finance their needs—but for years, the banks resisted. After all, those who had political power already had electricity. The development banks wanted to write big checks to national governments. You didn't get ahead at the World Bank by assembling a portfolio of small loans to benefit the least politically connected populations.

Still, some money flowed to the energy access sector. Cell phone companies in India and Africa, seeking both a cheaper way for them to power their cell towers and more reliable ways of charging phones for their customers, began investing in new business models for off-grid solar. At the same time, NGOs like Greenpeace, Oxfam, and the Sierra Club kept lobbying the banks to start lending money for solar power for the poor. Foreign aid agencies, tepidly at first, entered the field. America's private sector aid funder, the Overseas Private Investment Corporation, along with the Agency for International Development, created a clean energy financing facility for Africa, and demonstrated that with a government investment of only $40 million, it was possible to leverage $1 billion in private investment in clean energy. What the private sector needed was not subsidies but insurance.

Bangladesh took the business model, human capital, and finance challenge most seriously. Grameen Shakti, a spin-off incubated by the granddaddy of microfinance, Muhammad Yunus's Grameen Bank, pioneered rooftop solar for households. The government followed with the Infrastructure Development Company's Solar Home System Program, and by 2014 Bangladesh had three million home solar systems, second only to Germany.

In the summer of 2014, a year before the Paris Agreement, India's Congress Party was replaced by the BJP-led Modi government. Not long after, Modi pledged that every household would have at least a solar light by 2019. India was only at the starting line, but the race was on.

In 2016, when President Obama hosted Prime Minister Modi in Washington, a centerpiece of their meeting was the announcement of a partnership among the Indian government, the U.S. government, and American foundations to help provide early-stage funding to India's solar industry. Both the World Bank and the Asian Development Bank have committed substantial sums (in the form of low-cost loans) to finance energy for the poor. Electricity for India's villages, it seems, may finally have found some bankers. The challenge that remains is deploying the human resources—and expertise in building solar networks—necessary to provide electricity for India's 400 million energy-poor families. What India now needs is the kind of village organization that it has so often shown to be one of its great gifts. Gandhi should be smiling.

## SOLAR BATTLE ON THE ROOFTOPS

Meanwhile, in the United States, new players were taking advantage of both the staggering cheapness of solar panels and the increased cost of transmission on the U.S. grid to challenge established utilities. By offering to lease solar panels to homeowners on terms that guaranteed their electrical bills would go down from day one, companies like Sungevity,

Sunrun, and SolarCity installed more than a million rooftop solar systems from 2008 to 2015.

These rooftop systems generated surplus electrons—more than the homeowners needed—during the bright afternoon, the very time during which large-scale generators had charged the highest prices and made their profits. In leading rooftop solar states like California, the typical afternoon demand dwindled. Utility regulators began warning that unless something changed, the western United States would soon have a surplus of afternoon solar power. This would prevent utilities from profiting from their equivalent of surge pricing. From the perspective of the utilities, in other words, solar power had gotten too cheap, and they needed to slow it down.

A war broke out on America's rooftops. The fierce counterattack by utilities against the disruptive surge of rooftop solar peaked in the spring of 2016 on the West Coast. California's utilities asked regulators to stop requiring them to pay the retail price for electricity that homeowners fed into the grid and that the utilities would resell. The California Public Utilities Commission refused.

But next door, in Nevada, which has the highest percentage of solar energy of any state in the union, state regulators were more accommodating to the utilities. They sharply raised the fees on residential solar owners, while drastically cutting what the utility paid those customers for electricity. The new rules made rooftop solar no longer competitive in Nevada.

This was not a battle about solar versus fossil fuels. California's utilities long ago embraced renewables. Nevada Energy boasts some of the cheapest solar energy contracts in the nation. Utilities don't mind that solar is renewable, zero-carbon, and that its power is free—as long as they own it. But when solar is decentralized, it turns a utility's customers into competitors. Rooftop solar threatens the utilities' traditional rigid business model, with its guaranteed rate of return and monopoly on generation. By refusing to pay rooftop solar owners the value of the peak power they were generating, they hoped to guarantee that stream of profits—but by discouraging competition, the utilities raised afternoon power rates for everyone.

For society as a whole, this is merely one example of a core climate challenge: Cutting emissions requires accelerating the replacement of old, carbon-intensive technologies with new, cleaner ones. This replacement "strands" not only those who were paid to extract the coal and oil reserves that would be used to run power plants, factories, and refineries, but also those who profited from the plants, factories, and refineries themselves. The owners of those facilities don't want to find themselves with stranded assets, fighting to keep them operating.

For utilities, this transition to clean, localized power can only be managed if they compete in the new marketplace for rooftop solar. So far they have mostly refused, bitterly clinging to their old centralized model. But the *capo dei capi* of the private utility world, the Southern Company, has taken a different tack. When Georgia opened up its rooftop solar market, Southern launched its own branded rooftop solar platform. This laid the groundwork for Southern, taking advantage of its low cost of capital and trusted customer brand, to dominate the rooftop space rather than ceding it to newcomers like Sunrun or Sungevity. Southern CEO Tom Fanning made it clear in announcing this new venture that Southern had no intention of letting outsiders come into its historic service territory and poach the rooftop business unchallenged. (Fanning also likes to say of renewable electricity, "It's coming. You can't keep a wave off a beach.")

New York State has been providing leadership, too, negotiating a cease-fire between utilities and rooftop solar developers. The utilities make it easier for the developers to access customers and provide panels where the grid most needs them. In exchange, the developers agree to help the utilities pay for the grid.

The argument is sometimes made that, right now, most of the homeowners able to lease solar for their roofs are rich or at least well-off. That seems unfair—but it doesn't have to be. Lots of low-income families have flat roofs with good sunshine. Solar panels could lower their utility bills. Rooftop solar companies need the right kinds of regulatory support to be able to serve low-income as well as upper-middle-income communities. At the same time, utility regulators should be midwifing the rooftop revolution, not trying to abort it. Utilities need new rate structures and business models to compete in the rooftop space.

For all parties, then, it is business innovation as much as technological progress that will shape the pace at which rooftop solar realizes its promise.

## BUT THE SUN DOESN'T SHINE AT NIGHT

The biggest remaining barrier commonly cited to an all-clean, all-renewable, and zero-marginal-cost electricity supply is that solar and wind are intermittent resources: The sun and wind are not constant. This is really the last big argument that advocates of fossil fuel electricity can make.

Already in the United States, China, and India there is enough wind and solar in some regions that it goes to waste, because at certain times the grid can't absorb it all. In 2009, Texas was unable to use 17 percent of its wind power—there was no place for it to go. But it turns out that this seemingly enormous barrier has a simple solution. If, like most other nations, we had a national grid that could carry electrons swiftly and cheaply from regions where they are being generated to those in need, it would increase the reliability of our power supply, lower our bills, and enable us to rely on renewable energy like wind and solar for at least 80 percent of our electrical supply. In fact, on a smaller scale, Texas has built enough new connecting transmission lines that it now uses all the wind it generates—a lot more than it had back in 2009.

While sunshine reaches us only during the day (and not during cloudy periods), and wind in any one place is indeed fickle, it turns out that there is enough sunshine and wind across the United States at any given moment to meet our needs, if we can just get it to where it is needed. The solution to the intermittency of solar and wind is therefore connectivity.

Here, again, the answer to a climate challenge is to manage our affairs more efficiently. The United States is underinvesting in infrastructure. It routinely gets a D+ rating from the experts. The biggest deficit is in transportation. The second biggest is in the electricity sector: The American Society of Civil Engineers reports that a $900 million short-fall in investment in transmission and distribution will reduce national income by $1.8 billion each year.

Even these figures assume today's electricity mix. Tomorrow's system, in which the cheapest electricity comes from wind and solar, needs more, not less, transmission. In the summer, power demand in Chicago peaks after the sun begins to set. But at just about the same time, there is an early-afternoon solar generation peak in Arizona when solar farms generate excess power. If the United States, like China, built high-voltage direct current transmission (HVDC) lines from one region to another, those abundant Arizona afternoon electrons would arrive just in time to meet the needs of Chicago homeowners turning on their air conditioners after work.

Indeed, when scientists at the National Renewable Energy Laboratory researched the cheapest way to meet America's future electrical needs, they concluded that while building a network of HVDC transmission lines would add about 3 percent to the cost of providing electricity, it would enable the United States to access such abundant and cheap wind and solar that by 2030 the cost of electricity would fall by 10 percent—while the carbon dioxide released by each kilowatt hour (kWh) of power would fall by 80 percent.

Think of it this way: A grid with no coal, and 80 percent less carbon emissions, costs less than the one we will have if we just keep on doing business as usual.

## A NEW UTILITY BARGAIN

The accelerating decline in the cost of wind and solar will—on a cost basis—crush coal as a source of electricity. Yet we've now looked at three situations in which renewables are cheaper but still face substantial market barriers: energy access for the poor, rooftop solar, and grid connectivity. Price, it turns out, is a very big advantage to have—but not always enough. You also need adequate financing, business models that are fair to everyone, and incentives for transmission and connectivity.

We don't have these. The reason is simple. Electricity markets everywhere are based on early twentieth-century realities. Electricity, because of the need for wires to every house, appeared to be a natural monopoly. Electric utilities were given a bargain. They would be guaranteed profits—a fixed markup on their expenses—as long as they kept the lights on. Regulators would set both the allowable level of investment by utilities and the prices paid by customers. The prices had to be high enough to assure that the promised profits flowed but that there was always enough power. When regulators set prices too low, utilities lost money and typically failed to keep the lights on. Of course, regulators frequently didn't know enough to set prices and investment levels just right. This didn't matter for a long time, because two realities sustained the model:

1) Each generation of new technology was more electricity-dependent than its predecessor, so demand kept rising. Air conditioners needed more power than refrigerators.
2) Power plants kept getting bigger, more capital-intensive, and efficient—producing more electricity at lower cost per kWh.

So the utilities not only had a monopoly on the wire to your house, they also had no competition for the right to fill that wire with power.

If regulators approved wasteful capital investment, customers had to pay for the mistake—they were captive.

We need a new model, recognizing three new realities:

1) Demand for electricity from big power stations drops as efficiency and local generation rise.

2) Big nuclear and coal stations are getting more expensive per kWh; wind and solar are getting cheaper. Today, when utilities look around for "bigger is cheaper" options, they come up short. Most of their central station technologies—coal, nuclear, big hydro, new transmission corridors—are bad, unpopular neighbors. The bigger they are, the harder they are to site.

3) Customers are becoming empowered. Soon we will have millions of people generating some or all of their own electrons. Rooftop solar, cheaper than remotely delivered grid power when you include the cost of transmission, takes away the utility monopoly. It enables utility customers to declare independence.

While the research that led to these disruptive changes was in many cases inspired by concern over climate, renewable energy has now achieved a competitive advantage that is self-sustaining and no longer dependent upon climate policy. The sector will change dramatically—utilities cannot survive on their present models.

But the utilities are not down-and-out yet. They have enormous opportunities if they are willing to be revolutionaries, as they were in the 1920s when electrifying the nation, or in the 1950s when they enabled the post–World War II shift to an appliance-heavy, kilowatt-hour-dependent, suburban "all electric" life.

As IBM did a few years ago, utilities need to reinvent themselves. They need to deliver energy services, provide customers with flexibility and reliability, and welcome the idea that homeowners and businesses draw power from each other and from the grid. With this model, they would rely on fees for services customers chose, rather than being guaranteed a return on their investment. This is where most of the economy is heading—utilities need to catch up.

Today's public utilities would need permission from their regulators to convert from a capital rate–based monopoly to one reliant on more appropriate, modern mechanisms. For example, they could be rewarded for performance in delivering energy services chosen by customers. A diversified portfolio of fees could pay for transmission, with a remaining modest recovery from power generation. A utility would look more like a bank, which in today's world does not look like a bad profit strategy. If utilities insist on clinging to their outdated model for much longer, however, instead of leading the revolution, they will find themselves run over by it.

Utilities are not the only incumbents threatened by radical and rapid change in the electrical sector. Closing forty- to fifty-year-old polluting coal plants and replacing them with cheaper and cleaner renewables threatens the livelihoods of coal miners and utility workers, and can devastate the tax base of mining communities or small cities whose biggest taxpayer is a power plant. Owners, firms, communities, and workers are motivated to use every tool at their disposal to slow change to protect the value of their investments and livelihoods. The political uproar over the Obama administration's modest and long-overdue proposal to clean up grandfathered coal power plants is simply one example of the truth that profitable change can be blocked if we don't have the right alternatives available.

Every country has its own unique solution to the problem of regulating electrons, a commodity that cannot be stored and that has to be constantly available. So every nation will need its own set of reforms for a clean energy future. What is most important to understand is that in real economic terms, today's wind and solar electrons are the cheapest electrons the world has ever known. Those who develop them most intensively will make a great deal of money, and the challenge will be how to divide up the good news, not share the pain.

# PART IV

# GREEN LIVING

7

# WHERE WE LIVE

*You cannot have a national initiative
without involving cities.*
—Kasim Reed, mayor of Atlanta, Georgia

The Empire State Building is one of the most iconic buildings in the world. A 102-story engineering marvel when it opened in 1931, it was a towering symbol of New York's twentieth-century progress. Over time, however, it not only lost its distinction as the planet's tallest building (to the World Trade Center in the early 1970s), but it also became antiquated and far less efficient than its modern counterparts, leaking heat in the winter and cool air in the summer. This meant higher costs for both its tenants and owners.

Tony Malkin's family has operated the Empire State Building for decades. Eventually Malkin recognized that its iconic status only carried so much caché. The building needed modernizing to stay competitive. In 2008, he assembled a team to overhaul its heating, ventilation, air-conditioning, and lighting systems in order to meet the U.S. Green Building Council's "Gold" certification. Malkin's goal was to offer superior office space that, through energy efficiency, would cost less to operate.

Some of the changes his team implemented were small, like insulating all 6,514 glass windows and moving the thermostats in radiators away from exterior walls, which get hotter and colder than room temperatures (thereby triggering air-conditioning and heat more easily). Other changes included installing sensors for dimming or turning off lights, modernizing the refrigeration and exhaust systems, giving tenants the ability to monitor their energy use through individual metering, and optimizing office layouts to maximize natural light.

Malkin has made money going green. All told, the upgrades—called a "retrofit"—cost about $20 million, but the annual savings total more than $4 million. He has also proven that even in historical buildings, where renovation costs can be highest and most complicated, retrofits can pay dividends.

As we discussed earlier, cities are the source of about 70 percent of the world's greenhouse gas emissions. Buildings are the main culprit. They are responsible for consuming more than half the world's electricity, along with plenty of gas, oil, and HFCs to power boilers, air conditioners, and refrigerators. In addition, construction materials—cement, steel, plastic, glass, aluminum—are another major driver of emissions. At one point during China's construction boom, one-third of its carbon emissions were associated with making cement.

Making old buildings more efficient, and constructing new buildings to higher design standards, are an essential part of the battle against climate change. These changes hold substantial financial benefits, and so more and more major property owners are taking the same approach that Malkin did.

At Bloomberg, we have saved about $40 million over the past decade by reducing the emissions of our operations and information systems by nearly half. And, between our wind and solar investments, we obtain 23 percent of our electricity from renewable sources. By 2020, we aim to reach 35 percent, with a goal of becoming 100 percent renewable by 2025. The new European headquarters that we are building in London has its own power generation center that converts gas to electric power. We designed it to allow us to convert the waste heat to energy that can be used to warm the building in winter and chill it in summer. We've also tried to minimize the need for air-conditioning by designing the building with natural ventilation in mind.

Why would we go to all this trouble? For one thing, it's going to save us money. But let me give you a couple of other reasons. First, when you're trying to hire college graduates today, one of the questions they always ask you—and I come from a world where *you* ask *them*—is: What are you doing about climate change? Companies should have a good answer to that question if they want to attract the workforce they need. Second, now customers and investors regularly ask about companies' environmental policies. And if they don't like them, they may go elsewhere. That's less true of our business than of others, but we're seeing more and more of it. Many companies now advertise what they do for the environment, because it's good for their branding and their image. Action on climate change is now part of the demand curve, whether you like it or not.

Of course, not every green policy gets a lot of attention. Improving the efficiency of buildings is not as sexy as saving a rain forest. You don't see many celebrities dedicating their philanthropy to it. But the fact is, making the biggest possible dent in greenhouse gas emissions—and in the pollution that causes death and disease—requires focusing on buildings. And it's something that people can do in their own hometowns.

## BETTER BUILDINGS

To reduce emissions, one can't just tear down old buildings and put up new ones. In most places, that isn't possible, legal, desirable, or efficient. In New York, we worked in partnership with owners, landlords, and real estate developers to improve the energy efficiency of buildings in a number of different ways.

**Challenge Partners**. After we committed to reducing the carbon footprint of municipal buildings by 30 percent in ten years, we invited major property owners to join us in making the same pledge. Many did, and together they have saved more than $175 million on their energy bills while also significantly reducing their carbon emissions.

**White Roofs.** When roofs are dark in color, they absorb heat, which makes buildings warmer, requiring more energy to cool them. Light or reflective roofs, on the other hand, reflect heat and can dramatically lower the internal temperature of the top story of a building during hot days. That can save homeowners and tenants money and expand the life span of the roof—and all it takes is a special coating and a roller. In 2009, I joined former Vice President Al Gore on a rooftop in Queens to launch a volunteer program called NYC °Cool Roofs. The press poked fun at us for pushing paint rollers across the roof, but this simple fix is helping many New Yorkers save on their utility bills while also contributing to the fight against climate change. We also passed a bill requiring new roofs to be "cool." In a city with 1.6 billion square feet of roofs (the equivalent of almost 28,000 football fields), the savings, in both dollars and carbon emissions, are adding up.

**Clean Heat.** We always knew that the heating oil used in many residential buildings, #6 fuel oil, was not clean, but we never had good data that could tell us how much it contributed to air pollution. To get answers, in 2008 we placed street-level air quality monitors across the city—and the results were even worse than expected. The #6 oil was by far the largest local contributor to air pollution in the city. And it turned out that just 10,000 buildings—1 percent of the building stock—were responsible for the bulk of the problem, contributing more air pollution than all of New York City's cars and trucks combined. We considered simply limiting #6 oil, but given the harm it was causing to people and the environment, there was no way to justify its continued use, especially when there were cleaner alternatives available. So we banned it, starting in 2015. The real estate industry was not overly enthusiastic, but building owners also recognized that using more efficient fuels, while requiring some up-front capital spending, would save them (and their tenants) money in the long run. To help them finance the investments, we tapped into a federal program to create low-interest loans.

By following the data, making a relatively small change has led to a big impact on the city's air quality. In just three years, we lowered sulfur dioxide pollution by nearly 70 percent and soot levels by almost 25 percent. To put that in more meaningful terms: those reductions are saving around 800 lives a year and preventing 2,000 annual emergency room visits and hospitalizations. As a bonus, they are also reducing the city's greenhouse gas emissions by the equivalent of removing 160,000 cars from our streets. What's not to like?

**Building Codes.** Sometimes government can get in the way of green investments. In New York City, limits on buildings' height and size could sometimes deter retrofitting, because space is at such a premium. Building owners may not want to sacrifice even a few inches of space to thicken walls with insulation or a few feet of height to install a rooftop greenhouse. To free them to act, we exempted these kinds of actions from building and zoning restrictions. Now, a building owner can put solar panels or a turbine on the roof without it counting toward the height limit.

A big reason so many buildings remain inefficient is that landlords often don't pay the heating or electric bills; tenants do. So while tenants would prefer a more energy-efficient building with lower utility bills, landlords have little reason to make the capital improvements. As a result, buildings remain more costly to operate, and more carbon-intensive, than they need to be.

Fixing this problem starts with making it easier for landlords to collect data on energy use, in part so that tenants can use the data to shop around and choose energy efficient spaces. Regular energy audits of large buildings, which we required by law, now allow owners to learn ways in which their buildings were wasting energy. And the audits allow tenants to push for upgrades—or find less expensive space. We also required electricity sub-meters, which indicate how much energy each unit uses. In some buildings, tenants split the cost of electricity evenly. As a result, a tenant might feel no compunction about overusing the air-conditioning. It's like going out to dinner and knowing that others in your party are going to order the most expensive items on the menu, and you're all going to split the bill. If you have to pay for their indulgence, why shouldn't they pay for yours? Sub-metering is the equivalent of each diner getting his or her own check. Itemizing the building's electric costs incentivizes owners and tenants to conserve, which saves them money and reduces emissions.

Many of these changes were part of a package of laws called the Greener, Greater Buildings Plan, an overhaul of our building code focused on increasing energy efficiency. For the new building codes to be successfully implemented without getting tied up in the courts, we knew we needed support from the real estate industry. We sought it up front, by asking them to help us craft the codes in a way that they felt made sense and wouldn't be too intrusive.

Initially, building owners, and many unions representing building workers, were concerned that the retrofits would be too costly. But we made clear that our goal was achieving a certain level of energy efficiency, and that it would be up to buildings to decide how to do it. That changed the conversation. The building engineers told us that they could actually achieve a large share of the efficiency gains we were seeking by doing something called retro-commissioning. (With wonky names like that, no wonder celebrities don't get involved in this work.) Retro-commissioning a building basically means giving it a tune-up: making sure that the windows are caulked, hot water pipes are insulated, outlets aren't leaking power. These are simple, relatively cheap fixes that also create jobs for building workers.

In the end, the real estate industry and union leaders appreciated a government that was willing to listen to their concerns. Both groups supported our plan, even if it wasn't exactly what they wanted. When fully implemented, our building energy-efficiency policies will save $750 million a year and reduce emissions by 5 percent. The City Energy Project (which Bloomberg Philanthropies is supporting) is now applying the same energy-efficiency principles to twenty cities around the United States. It's amazing what can be accomplished when government, industry, and labor all work together.

## BUILDING MOMENTUM

City governments have led the way in cutting energy use in schools, hospitals, offices, and other public buildings. According to the Energy Information Agency, from 2003 to 2012, government buildings became more efficient by about a quarter, compared to commercial buildings, which improved by just 12 percent. There's a reason for this public sector edge. In order to make a building more energy-efficient, owners have to pay for the costly renovation up front. For many building owners, this proves challenging. Few have the capital. Interest rates can be prohibitive. And the period of repayment can be short, often only a few years.

In the same way that homeowners take out fifteen- and thirty-year mortgages to pay for homes they can't afford to buy outright, building owners should be able to finance sustainable investments. That has begun to happen through a program called Property Assessed Clean Energy (PACE), which allows building owners to pay off the value of their investments through an assessment on their property taxes. A private lender provides the capital, but because the loans are tied to the building's value, the risk is lower for lenders, and thus the interest rates are lower. Government collects the payments and transmits them back to the lender. If the owner defaults, the lender gains possession of the property.

Despite evidence showing the value of this approach, bank regulators have refused to let Fannie Mae and Freddie Mac underwrite loans on buildings in the PACE program. This threatened to be the program's death knell, but cities have stepped into the breach. Fresno, California, formed a partnership with a private financing firm that provides twenty-year financing with no up-front investment from the property owner. In Texas, the state legislature passed a law that gave local jurisdictions the power to develop their own PACE programs. As a result, Houston and other Texas cities now offer PACE financing for renewable energy projects. In fact, Houston, traditionally known as an oil town, has re-branded itself the "Energy Capital of the World." Its mayor, Sylvester Turner, proudly declares that the city "uses more renewable energy than any other city in the United States."

The same problem facing building owners—a lack of access to up-front capital—also confronts individuals who wish to convert their homes to solar energy. A whole industry has sprung up to help them afford solar panels. It works like this: homeowners essentially allow a solar provider to put panels on their roofs. In return, they purchase electricity from the panel provider at a set price—usually lower than the utility's price. One of the companies that supplies solar panels, SolarCity, is owned by Elon Musk, who also owns the electric automaker Tesla. His ultimate goal is not to put solar panels on top of the roof but to *replace* the roof itself with solar panels. Why pay for a shingled roof if a solar-paneled roof can work just as well—and power your home at the same time?

Of course, not every building can accommodate solar panels—for instance, they wouldn't work on top of Bloomberg's headquarters in New York City. But that doesn't preclude us from buying solar energy. We recently financed a solar energy installation on the roof of a warehouse in Queens. The energy gets fed into the grid, and while it doesn't go specifically to the Bloomberg building, we get the energy credits. It saves us about $28,000 per year. As companies develop innovative financing solutions like these, the financial barriers to sustainable buildings will continue to fall. Google has announced that it intends to use renewable energy credits to go carbon neutral by the end of 2017.

Technology is also changing the way we manage the use of power. Through mobile and connected devices (the so-called Internet of things), we're increasingly able to manage the energy use in our homes, offices, and stores. Soon, sensors will detect when rooms are unoccupied and adjust temperatures accordingly, saving energy from being wasted on heating or cooling empty places.

Even better, tomorrow's housing and building stock won't come with a utility bill. By combining high-tech materials and sophisticated planning and design, buildings can routinely generate enough electricity to power lights and appliances, and draw most of their heat and cooling from the ground underneath them or the sunlight falling on them. These "net zero" buildings will generate no greenhouse pollution. (The Bloomberg Center at the new Cornell Tech campus on Roosevelt Island in New York City, which is named in honor of my daughters Emma and Georgina, aims to be one of the nation's largest net-zero buildings.)

Buildings will have to use different techniques to accomplish this. A warehouse in Los Angeles may make extraordinary use of skylighting and rooftop solar, while a home in Germany might use ultra-efficient insulation and ground heat pumps. These buildings cost a little more up front, but they are cheaper to own and operate than traditional structures. The material and design revolution that enabled them is just beginning, and it will hold particularly big benefits for the developing world. That's where most new buildings will be constructed, where energy use is rapidly accelerating, and where the bulk of urbanization is occurring. In India, for example, a rising middle class is using more energy—for lights, computers, air conditioners, refrigerators, washing machines, and automobiles. The energy needed to meet projected air-conditioning growth in India alone may consume one to two times the electricity currently generated by all U.S. coal-fired power plants.

The trouble with all this extra energy is that the more we use, the more heat we generate. Think of the hood of a running car, or the bottom of a running laptop, or the surface of a lightbulb. Heat is a byproduct of energy use, and air conditioners are especially bad, because they take the heat from inside a building and dump it outside. The result can be enough to turn hot summer days into lethal heat waves. And the hotter it gets, the worse the problem becomes, because hotter days lead to more air conditioners working harder and longer.

But there are solutions. Look at China: twenty years ago, air-conditioning was rare in China—used by just one urban family in a hundred. Now, there are ninety-five air conditioners for every one hundred households. But partly because each tenant pays the bill, the Chinese typically set air conditioners to 77–80 degrees. Meanwhile, in India where electricity is heavily subsidized, households tend to set the thermostat lower. The best way to increase conservation is simple: make it financially rewarding.

We can also learn from the past. Much as I'd love to claim credit for white roofs as a New York idea, it wasn't. Cities in India long ago recognized the potential of white roofs to lower temperatures, as the historical city of Jaisalmer shows. (See photo insert.) Cities around India are rediscovering white rooftops, and the good news is that most of the buildings in India, Southeast Asia, and Africa that will exist in 2050 have yet to be built. This gives countries the opportunity to reverse the heat island effect, cooling their buildings from the outside in.

As new cities emerge, we have the advantage of knowing how, with smarter regulation and better technology, to make them climate-friendly and even carbon-neutral. Residing in a city is the most common way to live. What we do there has an outsize impact on our health and on our environment. If we're going to stop climate change, we must make smart investments in how we plan cities and how we design buildings.

In 2009, when I went to Copenhagen for the annual UN Climate Change Conference, I brought an eight-foot model of the Empire State Building and put it on display in the city center as a symbol of innovation. Discussing energy efficiency can be pretty dry and abstract. But when people heard that one of the planet's most recognized buildings went green in order to *make more money*, they had something tangible to remember. The Empire State Building, now approaching ninety years old, is still a symbol of the future.

8

# HOW WE EAT

*The reality is that the sustainable-food movement's reach will grow only to a point and ultimately will be limited to those with access, means, and education—unless legislators dramatically change food and agriculture policy.*
—ALICE WATERS, CHEF, CHEZ PANISSE

A confession: this was the chapter I procrastinated over the most. It's not because how we eat is not important—overall, food forms a major component of humanity's climate footprint. And not because the news from the food sector is doom and gloom—in fact, if we fail to reduce fossil fuel use fast enough and end up with intolerable levels of $CO_2$ in the atmosphere, a reformed agricultural sector can help come to our rescue, as I'll explain later. It's because one solution that figures prominently in the conversation about climate gives me a headache: changing our diet.

It is unequivocally true that a pound of feed-lot beef has many times the climate impact of a pound of equally nutritious beans, legumes, grains, or vegetables. An American eating a meat-heavy diet has a dietary climate impact twice that of an American vegetarian—3.3 tons of $CO_2$ per year versus 1.7 tons. A study by Oceana (which Bloomberg Philanthropies has supported) found that eating beef produces 4.5 times as much carbon emissions as eating wild-caught fish. These numbers have led to a heated debate over the role our diet plays in climate pollution. One key to climate progress, some environmentalists argue, is to kick our meat-eating habits and adopt some variety of vegetarian diet. Others have suggested economic incentives for changing our eating patterns. An Oxford University team calculated that taxing meat and milk 40 percent and 20 percent, respectively, would discourage consumption. These kinds of sumptuary taxes might have worked in the Middle Ages, but they are not viable climate-protection strategies in an increasingly democratic world. This is the climate change "sacrifice" idea on steroids.

Although there are many good reasons for eating less, or no, beef or meat in general, from my own experience I just don't see a massive dietary switch as a likely outcome of any conceivable climate campaign or program. Let me explain why with a personal story.

At the end of my Peace Corps service in Barhi, India, during the monsoon season, we had a small cholera outbreak. The health clinic watchman, Panditji, was one of the victims. In spite of being treated with intravenous fluids, which normally prevents cholera fatalities, he died. Our clinic doctor, Dr. Agarwala, was upset—he felt he had failed by losing a patient, particularly one of his own staff. Trying to reassure him, I pointed out, "Well, surely someone that old is more at risk if he does get cholera." "Old?" Dr. Agarwala asked. "Panditji wasn't old. He was only forty-five." I was shocked. The watchman had looked to be sixty-five or seventy. I told the doctor my mistake. "Oh, that happened during the famine three years ago," Dr. Agarwala said. "Panditji is a Brahmin, and very proud. He would only eat rice. Wheat and millet, which was all the grain available during the famine, were unacceptable to him. It was starvation during the famine, even though there was food, that aged him."

That story, to me, symbolizes how profoundly irrational and inflexible human dietary preferences are. We eat foods that reflect the tribe we belong to. What we grew up eating is likely to have a profound hold on our attitudes and tastes. So while I wish, for a variety of reasons including climate benefits, that McDonald's would lose its market appeal all over the world, and that healthier foods with lower methane foot-prints would replace the burger as a staple of modern living, I am not counting on a dietary revolution to save the East Antarctic ice sheet.

Having said that, we should keep in mind that smart eating (like walking instead of driving) is one of the only personal sacrifices that benefits the planet while simultaneously improving one's health. Arnold Schwarzenegger and filmmaker James Cameron (a vegan himself) have partnered with the Chinese Nutrition Society to urge the Chinese to eat less meat to protect the climate. Once again, what is good for our individual health is almost always good for the global climate.

How big of a deal is the total climate impact of eating? Big. Today, if you add up everything—growing crops, clearing and cultivating land, producing and applying fertilizers, shipping food to customers and keeping it refrigerated until it is eaten—food is responsible for about one-third of the total greenhouse emissions each year, from several different gases.

Agriculture generates 30 percent of total methane, mainly from livestock and rice paddies. Cows are the biggest source, because the way in which they turn grass into useful sugars involves methane-emitting bacteria in their rumen. Since there are 1.5 billion of them, belching and farting and excreting, their total methane is equivalent to almost 2 billion tons of $CO_2$, about equivalent to one-third of what the United States emits each year from its whole economy. Rice paddies generate another 700 million tons.

Farming also generates a greenhouse gas called nitrous oxide. Soil microbes create this gas when they digest nitrogen to provide it to growing plants, so the use of nitrogen fertilizer has greatly increased its scale. It is about 6 percent of the total climate burden created by the American economy.

Finally, modern agriculture has become a major source of $CO_2$. When natural ecosystems create soils, one of their main strategies is to capture $CO_2$ from the atmosphere and store it as solid carbon in the soil, turning it black. Prairie grasses are particularly effective at this trick. Across the world—in the American Midwest, Canada, Ukraine, China—there are huge fertile regions known as "black soil prairies" because of their enormous concentrations of carbon. For millennia, prairie grasses absorbed $CO_2$ to build their roots during the growing seasons. During the winter the roots rotted, and the carbon released was bound into the soil by microorganisms. But then came the plow, and industrial agriculture. Today 50–70 percent of this original soil carbon has been converted from solid to gaseous $CO_2$ by poor agricultural practices.

In addition, a lot of the industrial activities that support modern agriculture—driving tractors, producing and spraying pesticides, mechanized planting and harvesting of crops, shipping produce to market, refrigerating food, and disposing of agricultural and food wastes after the harvest or meal—contribute in varying degrees to $CO_2$, methane, halocarbons, and other greenhouse gases.

And there's one more way that food affects the climate: cooking.

About three billion people in developing nations lack access to modern cooking fuels, relying instead on biomass: wood, straw, cow dung, or charcoal. Many burn these materials on three-stone fires—just a pit with three stones to hold up a cooking pot. Others use clay stoves. Both turn a large part of the fuel not into heat, but black carbon or soot. The soot produced by biomass cooking is estimated to kill 4 million people every year—more than malaria or tuberculosis. Most are women and children. Cooking—preparing food—is estimated by the World Health Organization to be the fourth largest risk factor for disease in the developing world. Huge efforts have been made over the past thirty years to create an affordable stove that will burn the same primitive fuels more cleanly and efficiently. Engineers can make these stoves, but none to date have proven satisfactory to the women who are the ultimate customers.

The black carbon emitted by biomass cooking is a huge climate problem—as is the deforestation that results from harvesting the wood for charcoal or burning. Primitive cooking generates about 25 percent of the world's black carbon, so it's about 4 percent of the total climate problem. Women without access to modern fuels spend one to five hours a day gathering fuel, so the health benefits of clean cooking are staggering.

Currently, too little progress is being made on providing access to clean cooking, because there is no magic technology solution like solar to drive the market into remote and poor villages. Yet that doesn't mean we don't have solutions. Success, unsurprisingly, looks a lot like prosperity: If families can afford propane, LPG, or liquid ethanol, the stoves that burn these fuels are affordable. Here is one place that government subsidies could prove critical, and where offsets and other forms of carbon finance might be powerful development tools. The next generation of women in Africa and South Asia should not be forced to cook over toxic stoves, but instead have access to the same fuels that the middle class enjoys and that are standard throughout most of Europe and North America—clean liquid or gaseous fuels like natural gas, propane, and ethanol.

## THE FUTURE OF FOOD

Eating is a big deal for the climate. And it's complicated. But it's also a source of hope, because there are so many things that so many people and institutions can easily do to make agriculture a climate solution, not part of the climate problem. Beyond changes in the food we eat, there remain lots of reforms and interventions that we can undertake. Our present way of growing, transporting, storing, and cooking what we eat is wasteful in ways that do not titillate our taste buds, nourish our bodies, or remind us of Grandma.

Let's begin with a no-brainer.

About one-fifth to one-third of all food grown is wasted: either between the farm and the table, through spoilage; or after dinner, as garbage. Investments in better food processing and handling systems, particularly in poor countries, would cut those losses dramatically, improving farming incomes and reducing poverty.

In the developing world, small farmers lose enormous parts of their income to pests and spoilage. A simple three-plastic-bag solution developed at Purdue University can keep grain fresh for months. Small metal storage silos can eliminate food loss during storage. Even substituting a plastic crate for a bag during transportation can dramatically cut losses of fruits and vegetables from bruising, which can render them unfit for sale. One farmer in Florida had to dump 24,000 pounds of spaghetti squash for the simple reason that it had wind-scoring marks on the rind, and thus no retailer wanted to buy it—even when he offered it for sale at a steep discount. We waste one-third to one-half of all the fruits and vegetables we produce. A desire for standardized, cosmetic produce for marketing purposes drives much of this loss. Eliminating food "quality" standards that are simply designed to make produce look prettier would help.

Even food that gets sold doesn't always end up on the table. Americans throw an appalling amount of produce away each year—about 50 million tons (in addition to the 10 million tons that rot in the field). That's worth more than 1 percent of our total GDP—$218 billion, about nine times as much as our government spends on humanitarian foreign aid. A City-Lab study concluded that a lot of waste could be headed off by standardizing the way in which expiration labels are presented to customers.

Another source of food waste—and overeating—is container size. Cutting the size of a plate from twelve inches to ten inches causes diners to eat up to 22 percent less food. The amount of food in a precooked portion also influences how much we eat. Manufacturers know this and have been increasing portion size as a marketing ploy. In Britain, for example, frozen chicken pies have gotten 40 percent bigger. Mike Bloomberg's effort to curb the size of soft drinks sold in New York would have had climate, as well as health, benefits, by reducing both waste and waistlines.

I've indicated my skepticism about massive changes in dietary preferences as a likely climate strategy. But smaller tweaks like the ones I've just outlined can make a surprisingly large difference. And there are more.

Six ounces of chicken have one-tenth the climate impact of the same serving size of Brazilian beef. An American who doesn't eat beef but eats other meats has a climate footprint only 10 percent higher than a U.S. vegetarian, and 20 percent higher than a vegan. Vegans do make a smaller impact on the climate than carnivores. But a vegetarian diet heavy on dairy might have a bigger impact than a pescatarian diet. The World Resources Institute has prepared a protein scorecard if you want to examine your own diet and figure out if you want to make changes.

Fish are generally good in terms of their climate impact, but not all seafood is climate-friendly. This is especially true of shrimp farmed along coastlines, which often comes at the expense of carbon-storing and coastline-protecting mangrove forests. Far more work is needed on projects like the Mangroves and Markets effort in Cà Mau, Vietnam, to find ways to intensify shrimp production from the natural mangrove habitat without destroying it.

Even if we streamline our food supply chain and eliminate avoidable waste, at the end of the meal there will always be garbage. How we handle it right now, however, is a major scandal: Generating about a tenth of man-made methane, food waste is a big and avoidable part of the climate impact. If we expose it to enough oxygen for aerobic bacteria to break down, food waste will naturally compost into organic soil, retaining the bulk of its original carbon. Tragically our so-called "sanitary landfills," into which most cities send their unsorted waste, interfere with this natural process by sealing garbage underground with no oxygen. This is the perfect formula for turning it not into compost but methane, which then escapes and heats the climate. Even when landfill operators try to capture the methane from decaying garbage and use it to generate electricity, or as biogas for operating garbage trucks (as Los Angeles does), most of the methane created in the landfill still escapes as pollution.

The solution? Keep garbage separate from recyclable paper and metal. Use landfills only to dispose of inorganic materials like broken bricks. Everything else becomes compost, biogas, or is recycled. The landfills that already dot the landscape will continue to be a methane problem for a few more decades, but we should start separating garbage from other waste immediately and abandon the perverse model of the sanitary land-fill—which could not have been better designed to mess up the climate.

Some of the countries with the worst solid waste problems today—in South Asia and Africa—have an enormous opportunity to build cities of the future with the right kind of solid waste handling systems, so that waste as a major source of methane becomes a memory.

## HOW WE GROW

Agricultural practices are also critical. The way farmers grow their crops makes a huge difference in their climate impact. Overuse of nitrogen fertilizer, excessive plowing, failure to plant winter cover crops, cultivating steep hillsides instead of using them for pasture, allowing erosion of streamside vegetation zones—all such practices exacerbate the impact of growing a meal. One European study suggested that revising such practices could cut both methane and CO2 emissions in the EU by 50 percent.

Fertilizer use is one of the most critical contributors to climate risk. Nitrogen fertilizers are the primary source of the third most damaging greenhouse gas, nitrous oxide. Its concentration in the atmosphere has increased by 20 percent since the Industrial Revolution kicked off, mostly in the last fifty years as synthetic fertilizer use soared.

Because nitrous oxide also comes from other sources—natural soil microbes, ocean plants, and rain forest destruction—until recently there was no way to measure how much was coming from fertilizer, and there-fore little pressure on agribusiness to do much—another example of

Mike's adage that you can't manage what you don't measure. Since 2012, however, we have been able to tell whether a molecule of N2O came from a natural source or fertilizer, and the data reveals that it is fertilizer overuse that has driven up its concentration in the atmosphere.

Since nitrous oxide lasts 120 years, it's urgent that we try to curb it quickly. (Nitrogen fertilizer is also a major source of soil carbon stripping, discussed below.) Ironically, a huge proportion of today's fertilizer use has no benefits to the farmer. When fertilizer is applied just before a rainfall, for example, most of it washes off into streams, eventually contaminating drinking water and creating oceanic dead zones at the mouths of major rivers like the Mississippi. We can't stop using fertilizer at once, and in some situations we may always need it. But there are still reforms we can put into place immediately.

In today's industrial agriculture, farmers traditionally overuse fertilizer, applying it equally to all parts of their fields, no matter where it is needed most and even where it isn't needed at all. Fortunately, new precision technologies are increasingly enabling industrial producers to use the exact amount of fertilizer needed, and new formulations are enabling nitrogen to be delivered only as fast as plants can utilize it.

In much of the world, however, government funds have gone into subsidizing the cost of fertilizer rather than into enabling farmers to use it more effectively. In other instances, the high costs of potassium and phospate, which are customarily paired with nitrogen fertilizer, has led to unintended consequences. In India, for instance, nitrogen is heavily subsidized. But for nitrogen to be effective, it must be combined with phosphate and potassium—which are not subsidized and quite expensive. As a result, many Indian farmers wastefully apply nitrogen to their fields. By some estimates, half of the total nitrogen used is simply converted into either air or water pollution, because it is paired with too little phosphate and potassium.

Admittedly, even simple steps, like eliminating fertilizer subsidies that actually cost farmers income, involve reeducation, reforms, and resistance. At the same time, both for economic and environmental reasons, this misuse and waste of nitrogen is being challenged. Ethiopia equipped its farmers with a digital map of soil productivity through a nationwide Soil Information Service, giving them much more precise information about the fertilizer needs of their soils, and thereby tripling wheat yields in 40 percent of its croplands.

Farm practices can also do something about agricultural methane. Some breeds of cows, for example, release far more methane than others per gallon of milk produced. Rice paddies can be managed better as well. In Asian, African, and Latin American rice-growing regions, farmers are increasingly shifting to alternative wetting and drying (AWD) irrigation practices, in which the rice field is periodically allowed to dry out during the growing season. Because methane from rice paddies is only produced when vegetation in the field rots anaerobically, flooding just the furrow (not the entire field), and only when water is *needed*, cuts methane generation by 90 percent. These practices also save a farmer—and the region—water.

One of the biggest available climate solutions is to end, and then reverse, the stripping of carbon from the soil by abusive agricultural practices. In the United States, China, India, and Latin America, prairie soils have been denuded of billions of tons of carbon by being plowed deeply (leaving bare soil exposed during the fallow season and winter) and then doused with artificial fertilizers. Plowing and bare soils encourage carbon in the soil to be converted to $CO_2$ through oxidation. The soil microorganisms (fungae, bacteria) that carry out the trick of locking carbon into the soil had never encountered—and cannot survive— repeated overgrazing, erosive plowing, and huge concentrations of nitrogen fertilizers.

Similar soil degradation has occurred wherever industrial agriculture has taken over the landscape to grow wheat, corn, soybeans, cotton, or other grains with heavy inputs of fertilizer and pesticides. As industrial agriculture, and particularly the fertilizer revolutions, spread across the world's farming belts, formerly black soils began emitting their stored carbon as greenhouse pollution. They turned red.

Do we have alternative ways to grow our food? In 1981, as concern about climate change was nudging over the horizon, the Rodale Institute in Kutztown, Pennsylvania, initiated a Farming Systems Trial to attempt to answer this question. What would happen to soil carbon content if corn and soybean farms were converted from conventional industrial U.S. agriculture to what Rodale dubbed "regenerative agriculture"? Rodale's regenerative agriculture avoided chemical fertilizers and pesticides, replacing them with conservation tillage, cover crops, rigorous crop rotation, composting, and residue retention.

The results were inspiring. By 2005, the tests had turned croplands that had steadily been carbon emitters or sources into carbon sinks. Not only that, the regeneration system also cost less, used less water, generated higher profits, and produced equal yields! Every year, an acre of regeneratively farmed cropland converted 3.5 tons of atmospheric $CO_2$ into stable, fertility-enhancing soil carbon. Tropical tests stored even more, and better management of grasslands turned out to be the most potent potential strategy of all for reducing atmospheric $CO_2$.

Rodale is not the only expert voice that believes reforming agricultural practices can reverse the accumulation of $CO_2$ concentrations in the atmosphere. Ohio State University agronomist Ratan Lal believes that simply focusing on restoring degraded and desertified soils could store 3.5–11 billion tons of $CO_2$ annually. Even the agribusiness giant Monsanto, which disagrees with Rodale on almost everything, reports that strategies like cover crops could greatly increase carbon storage in soils, particularly in corn and wheat regions.

This is a critical finding. Many scientists fear that we will increase $CO_2$ concentrations so much that we will face devastating risks—risks that we can only avoid if we can somehow reduce greenhouse gas concentrations. We'll discuss this more later, but if we do enter the danger zone, it is plants—billions and billions of them—that have the best shot at pulling the excess CO2 out of the atmosphere and restoring climatic stability.

So farming—along with eating and cooking—offers some of the most promising climate solutions with some of the most appealing immediate benefits: food security, healthier diets, less waste, greener landscapes, and more reliable water supplies. But the most important thing to remember about farming is that it was plants that originally created the gentle Holocene climate that nurtured civilization. If we discover that we have threatened that civilization by overloading the atmosphere with greenhouse pollution, it is plants that will most likely bail us out.

# PART V

# TRAVEL DIRECTIONS

9

# CITIES TAKE THE WHEEL

*As urban populations continue to grow, we cannot rely*
*on the business-as-usual scenario of car-based cities.*
—Yeom Tae-young, mayor of Suwon, South Korea

In 1992, the UN named Mexico City the most polluted city on the planet. Its toxic air took on mythic proportions—birds were dying midflight over the city! Or so people said. Whether it was true didn't matter, because it was believable. The air was that bad. The biggest cause of all the pollution? Traffic.

The pollution was devastating for public health, causing an estimated 1,000 premature deaths and 35,000 hospitalizations each year. It also had a terrible impact on the city's business environment. Cities that hope to attract international investment need to create environments that allow companies to draw the talent they need to succeed. For any CEO of a business that employs skilled labor, talent matters more than tax rates in determining where to invest. People want to live in places where their well-being is not threatened.

To their credit, Mexico City and the Mexican federal government recognized the problem and took action. They imposed stricter fuel emissions standards on cars and began restricting driving on alternating days, depending on whether a vehicle's license plate ended in an odd or even number. But the system didn't work as intended, for the same reasons it hasn't worked everywhere else it's been tried. Those who can afford to do so simply buy a second car—often an old clunker without modern emissions controls—and the second car has a different plate. The working and middle classes have a harder time getting to work, and the air typically remains filthy. No one benefits.

Mexico City eventually adopted more effective strategies, by investing in its subway and bus systems, switching to lower-emissions buses, and starting public bike-sharing systems. While pollution and congestion remain problems, the air is significantly cleaner than it was in the early 1990s. That has saved lives and strengthened the city's business climate. The steps Mexico City took to reduce air pollution also shrank its carbon footprint. In 2008, the city set a goal of reducing carbon emissions by 7 million tons by 2012—the equivalent of the total emissions from around one million homes. It exceeded its goal by more than 10 percent. By far the biggest reductions were from transportation. Today, Mexico City—the former world #1—has fallen far down the list of the most polluted cities.

Mexico City shows that progress is possible with strong leadership and smart policies, and the city is working to build on its air quality gains. But there's another reason why Mexico City has dropped in the rankings of polluted cities that is less encouraging: Other cities around the world are getting worse, thanks in part to increasing traffic congestion.


The invention of the automobile revolutionized the way we lived, improving it in immeasurable ways. But it also changed the way political leaders and urban planners thought about cities, often with destructive results. As more and more people were able to afford cars, more and more space was given over to accommodate them. Trains were the past. Cars were the future. And when roads became plagued with ever-increasing traffic, the answer was to double down, by building bigger roads with more lanes. Inevitably, those became choked with traffic, too.

Over time, automobiles became a major source of greenhouse gas pollution. Transportation, of both people and goods, creates about 14 percent of global climate emissions. Almost all of it is from oil, which releases CO2 when burned. In 2016, climate emissions from oil in Europe and the United States exceeded those from coal for the first time. In many of the world's cities, transportation is the single largest source of greenhouse gases.

Cars remain essential to the way we live. Nevertheless, for good reasons, cities have begun rethinking their relationship with cars. Cities have strong incentives—having nothing to do with climate change—to reduce the number of cars on the road. Traffic congestion harms their economies, health, and safety.

**Economy.** In 2014, American city dwellers spent 6.9 billion extra hours driving because of traffic congestion, costing the economy more than $160 billion in higher costs and lost productivity—to say nothing of the frustration of being stuck in traffic.

**Health.** Particulate matter from car exhaust causes cancer, asthma, cardiovascular disease, and other serious respiratory problems, making it a significant contributor to the more than 7 million annual premature deaths globally from air pollution.

**Safety.** Each year, more than 1.25 million people die as a result of car crashes, many of them in cities. Tens of millions more are injured. As cities continue to grow, traffic will claim more and more lives, unless we do something about it. The number of cars on the road surpassed 1 billion in 2010. By 2035, that number could double.

China is the number one driver of the growth in automobiles, and that is one reason why air pollution has become an urgent crisis there. According to the Health Effects Institute, 137,000 premature deaths each year in China are attributable to transportation pollution. But China is far from alone. Beijing, for all the news photographs of impenetrable smog, is actually ranked fifty-seventh among the world's cities in air pollution.

Traffic isn't the only source of pollution in those cities, but it is a major one. In Indonesia, where rising incomes have put more cars on the road, a study showed that 58 percent of all illnesses among residents of Jakarta were related to air pollution in 2011. According to the nation's health ministry, between 70 and 80 percent of air pollution in Jakarta comes from vehicles. More cars lead to more pollution and more sickness.

This is not only a crisis in developing countries. On March 18, 2015, thanks to a combination of weather conditions that worsened smog— and despite the bold climate agenda recently initiated by the city's new mayor, Anne Hidalgo—Paris briefly had the most polluted air of any city in the world. (As Carl explains in the next chapter, the so-called clean diesel cars that EU countries have been favoring aren't so clean.) The City of Light was dimmed by a toxic cloud, and Mayor Hidalgo swung into action. Public transit was made temporarily free. A partial ban on driving was instituted.

After the emergency passed, the city put in place more long-term solutions. Cars made before 1997, when Europe instituted its first emissions standards, are banned during the busiest hours on weekdays. By 2020, only cars built after 2011 will be allowed to enter Paris. Plans are in place to eventually ban all diesel vehicles.

## STREET DREAMS

For a long time, conventional wisdom held that the needs of people were one and the same with the needs of cars. But that was never true. And today, more city leaders are recognizing that when the interests of cars and people diverge, people should come first. That shouldn't be a revolutionary concept, but sometimes the biggest changes revolve around the simplest ideas.

In 2009, my transportation commissioner, Janette Sadik-Khan, came to me with a proposal: close Times Square to vehicular traffic. I thought she was joking. She wasn't. Times Square? The Crossroads of the World? The home of Broadway? One of the busiest intersections on the planet? *That* Times Square? I told her she was out of her mind. (And I may have used a certain adjective before the word "mind.")

My gut reaction—and nearly everyone else's, I think it's fair to say—was to think that closing streets would lead to more traffic congestion. But as I listened to Janette, I began to see the issue differently. Traffic in the area had been a nightmare for decades, which seemed largely inevitable. After all, Times Square is one of the most popular tourist destinations anywhere in the world. It sits at the epicenter of the world's economic capital. Millions of tourists visit every year, and millions more travel through Times Square to get to work. Congested? Of course it's congested.

Manhattan's streets are laid out in a grid, which makes it easy for tourists to find their way around without getting lost (until they enter the West Village). City fathers adopted the grid in 1811 to bring more order to the city's development, but one major thoroughfare that did not conform to the grid—Broadway—was left intact. Broadway, which runs from Lower Manhattan to the northern tip of the island, follows a path blazed centuries earlier by Native Americans. It runs on a diagonal between perpendicular avenues, creating triangles of space each time it cuts across a new avenue. These triangles have been dubbed squares: Greeley Square, Madison Square, Herald Square, Times Square, and Lincoln Square. And each is a place where automobiles, pedestrians, cyclists, shoppers, hot dog vendors, and people from every imaginable walk of life compete for space.

At Times Square, unsurprisingly, the competition was most intense.

Throngs of people, many of them gawking at the bright lights above them, ignored sidewalks and walked in the street, because there was often no room on the sidewalks. Collisions with cars were as dangerous as they were common: The number of pedestrians hit by cars in Times Square was more than double the number on surrounding avenues.

But Janette presented the counterintuitive case that closing streets could actually speed up traffic. She believed, after looking at the issue with the city's traffic engineers, that closing Broadway to cars between some of those angled crossings could actually improve traffic flow by reducing the number of intersections. With a little rejiggering, she told me, we could give drivers through Times Square more overall green-light time, while also giving pedestrians another two and a half acres of public space at the center of the world.

It was plausible enough to give it a try. And if it failed? Well, live and learn. I didn't run for mayor to play it safe. Some of my political advisers may have wondered why it couldn't wait until after my reelection campaign was over, but they knew better than to try to talk me out of it.

Janette and I announced a six-month pilot project for pedestrian plazas in Times Square and Herald Square. I guessed that many people would have the same initial reaction that I did: that we were nuts. Judging from the immediate reaction in the press, I wasn't wrong—I just underestimated. One columnist for the *New York Post* screamed that Times Square would soon be known as the "Traffic-Choked, Tourist-Loving, New Yorker-Hating, Immovable Crosswalk that Mayor Bloomberg has diabolically envisioned for the middle of this town." She called the plaza idea "downright mean." A column titled

## B'WAY BLOOMY'S ROAD-KILL PLAN

was followed a couple of days later by another titled

## DEAD END STREETS.

For the opening of the plaza, Janette had ordered metal chairs and tables, but because of a hiccup they didn't arrive in time. So she improvised a solution, finding a bargain deal for hundreds of beach chairs in rainbow colors from a hardware store in Brooklyn. As soon as they were set up, people sat down in them as though it was the most natural thing in the world to do. At that moment, we knew the plaza would be a success—and it was. The plaza filled with people enjoying a new experience of Times Square, and it has stayed that way ever since. And in classic New York fashion, many of the same people who called us insane for proposing the idea in the first place now complained, "Why can't we have this kind of plaza in my neighborhood?"

But before we expanded the idea to other areas, we first had to answer a crucial question: What impact would it have on traffic?

For the first hundred years of the automobile's existence, cities had to measure traffic with a watch. Transportation departments would send drivers to cruise (or crawl, as was often the case) through an area over and over again to measure their speed. This approach invites inaccuracy. Today we have a far better tool at our disposal: GPS. By 2009, we had already required that each of the city's 13,000 yellow cabs have the system. And because so many cabs pass through Times Square each day, we already had a treasure trove of data on their speed—both before and after the pedestrian plaza was created.

Data from 1.1 million taxi trips showed that traffic times around Times Square improved by 7 percent after the redesign. Most important, the number of people injured in crashes fell by 35 percent, mostly because pedestrians and automobiles were no longer vying for the same space. Before the change, 89 percent of the area in the heart of Times Square was dedicated to vehicular traffic, even though 82 percent of the people passing through Times Square every day did so on foot. By creating the plaza, we helped balance the scales for pedestrians.

Businesses around the plazas had initially worried that the new arrangements would hurt their bottom line. Just the opposite happened.

By 2013, retail rents along the Times Square plaza had doubled, and for the first time in its storied history, the area became one of the top ten most valuable retail destinations in the world.

The Times Square pedestrian plaza exceeded our highest expectations, and we took the model citywide. We built more than fifty other pedestrian plazas, which freed up dozens of acres of new space for people to use. Each brought the same kind of benefits to neighborhoods that closing Broadway had brought to Times Square: more customers for businesses, safer streets for pedestrians, and cleaner air for all.

To chip away at car traffic and carbon footprints, and to rebalance the scales for pedestrians and cyclists, cities around the world are rethinking the way streets are used. Paris transformed part of the Left Bank of the Seine into a car-free pedestrian zone in 2013, and plans are under way to convert a section of the Right Bank from the Place de la Bastille to the Louvre—a stretch over which up to 2,700 cars pass per hour—into a walkable park. Madrid, Spain, is slowly expanding the car-free zone it created in 2015, with the goal of making the city center essentially car-free by 2020. Barcelona is creating what it calls "superblocks": conglomerations of multiple city blocks where there is no through traffic. Intersections are reclaimed as public spaces, and streets are returned to pedestrians and cyclists. Because it's simple to adapt, the idea has captured the attention of cities around the world—and if superblocks prove a success, it's easy to imagine them spreading from city to city.

Other cities are going further. Chengdu, China, is constructing a satellite city that largely excludes cars—one of many cities that is developing long-term plans to redesign neighborhoods around walking, biking, and mass transit. Bogotá, Colombia, has been a forward-thinking city on transportation, thanks to its dynamic mayor, Enrique Peñalosa, who says: "We have made a city much more for people and less for cars. I took tens of thousands of [parked] cars off the sidewalks and we made new sidewalks.

We had TV commercials explaining sidewalks are for talking, for playing, for doing business, for kissing. We've started a sidewalk revolution. TransMilenio, the BRT (bus rapid transit) system, was also a very powerful equality symbol because we took space away from cars to give it to public transport. And for the first time we had the people in public transport going faster than those in cars. It shows there is democracy; it shows all citizens are equal. We created the Alameda El Porvenir—maybe the achievement I am most proud of—which is a bicycle highway 50 meters wide and 24 kilometers long that thousands of people use every day to go to work."

Here in the United States, cities from every region of the country—big and small, some led by Democrats, others by Republicans—are embracing these kinds of reforms. Take Jim Brainard, the Republican mayor of Carmel, Indiana: "We have examined every area of city government, from adding hybrid cars, investing in solar programs, creating bicycle and pedestrian facilities, and designing the city for people, not cars, with the goal of making our city as environmentally friendly as possible." Or William Bell, the Democratic mayor of Birmingham, Alabama: "This year we've rolled out a new bike-share program, introduced electric vehicle charging stations, and more than doubled the number of vehicles running on alternative fuel. We're also going to transition our remaining fleet of gas-powered buses to 100 percent electric. These actions help our residents, cutting gas bills and improving health. They also show leaders on the national and international stage that change is possible, desired, and beneficial."

Portland, Oregon, has an ambitious vision based on a measurable goal: making it possible by 2030 for 90 percent of Portlanders to meet all their non-work needs either on foot or on bike. Even Detroit—Motor City—is exploring the concept of "20-minute neighborhoods," where residents can meet all of their non-work needs within a twenty-minute walk or bike ride.

In New York City, we transitioned the city's fleet of 13,000 taxis to hybrid and other high-mileage vehicles, increasing fuel efficiency by about 50 percent. And we adapted many ideas from other cities, including bike lanes in Copenhagen and Amsterdam. The last time a New York mayor had built bike lanes, in the 1980s, the experiment didn't work—largely because too few lanes were built. We built more than 470 miles of bike lanes in twelve years, including protected bike lanes that are separated from car traffic by parking spaces. These steps helped to upend long-held assumptions that biking the streets of New York was only for delivery workers and cycling fanatics.

Bike ridership in New York, and in North American cities in general, has historically been fairly low. At around 7 percent, Portland has America's highest rate of people cycling to work. (In Amsterdam, by contrast, nearly 40 percent of all trips are on bicycles.) For years, many people took it for granted that New York would never be a biking town. But make streets safer, and more people will pedal. From 2001 to 2012 the number of bike commuters quadrupled—without any increase in the number of serious bike crashes annually. On First and Second Avenues, protected bike lanes increased the number of bikes by 177 percent. On Ninth Avenue, where we'd built the city's first protected bike lane, injuries to all street users fell by 58 percent, even as bike ridership soared. The better the infrastructure, the more ridership will grow. Copenhagen and Amsterdam weren't always world capitals of cycling—they had to make streets bike-friendly first. More cities are catching on.

## SHARING THE ROAD

In May 2013, we launched New York's Citi Bike system, the largest public bike-sharing program in the United States. We had installed more than 6,000 bikes at stations around the city and created a new public transportation system—at zero cost to taxpayers, thanks to a public-private partnership with the financial firm Citi, which paid for the system in exchange for branding and a cut of long-term profits.

In developing our program, we learned from the pioneering work done by cities like Paris, whose enormously successful Vélib' system has inspired cities around the world. Many people doubted that a similar system could work in New York. But with no public money at stake, I was willing to give it a shot. I'm glad I did. A year after we launched the program, total distance traveled on Citi Bikes passed the 20-million-mile mark—and neighborhoods around the city were clamoring for stations. Today, there are more than 8,000 Citi Bikes, and counting.

Ride sharing—bike and automobile—is one of the most exciting frontiers in urban transportation. Building on the great success of Vélib', and recognizing the potential for sharing another low-carbon option, Paris launched a subscription-based electric-vehicle (EV) sharing system called Autolib'. The city has started an electric scooter–sharing service, too. Singapore is creating an EV-share program that will include a fleet of 1,000 electric cars and 500 charging stations by 2020. Indianapolis's BlueIndy program includes plans for 500 electric cars and 200 charging stations. By connecting on-demand electric car rides with the Indiana Pacers Bikeshare program, which operates in the city's downtown, more and more residents of the car-racing capital of the world can get around town on their own schedule with zero carbon emissions—and without having to own a car.

Southeast of Denver, the city of Centennial, Colorado, created a partnership with the ride-hailing company Lyft to encourage more commuters to use public transit. Under the agreement, the city will subsidize Lyft rides to and from a train station where commuters can catch trains to the city center. The cost of subsidizing the rides is less than Centennial was paying for its existing dial-a-ride service, so the city saves money and fewer people drive into the city center. Or at least that's the hope. In a survey done before the launch, 90 percent of Centennial commuters drove to work in Denver—most of them because of speed and convenience. With a better alternative, more residents of Centennial may well leave their cars at home. It's worth a shot.

## HOP ON THE BUS (AND THE TRAM)

Taking streets back from cars requires giving people reliable, affordable alternatives. Underground and overground urban rail systems provide the greatest mobility to the greatest number of people. They also emit the least carbon and air pollution. Still, they're expensive to build. Bus systems, on the other hand, are far cheaper and far easier to operate, maintain, and expand. The only trouble is that they tend to be slower. This adds to traffic and air pollution, not to mention travel times. It also makes them less attractive to commuters, many of whom decide to drive cars instead—thereby adding further to the traffic and air pollution.

In the 1970s, the Brazilian city of Curitiba thought of a way around this: make buses work more like trains. The city gradually introduced a number of innovations to its bus system. Riders prepaid at roadside kiosks and could board at multiple points on the bus, which eliminated lines and sped up travel time. Special lanes were set aside, enabling the use of longer, higher-capacity buses, which increased the system's capacity.

The idea, which has come to be known as bus rapid transit, at first caught on in only a few other cities, but in recent years it has taken off. There are now bus rapid transit systems in more than 200 cities around the world, carrying more than 33 million passengers every day, more than 20 million of them in Latin America.

Bus rapid transit is a great example of how transit improvements can be effective not only for moving people out of cars but also for moving them out of poverty. In Johannesburg, as in many cities around the world, some of the poorest residents live on the outskirts of the city, farthest from jobs and opportunities. It therefore takes longer—and often costs more money—for people in poor communities to get to work, and for their children to get to school. In Johannesburg, this disparity is in part a legacy of apartheid.

When I attended a C40 conference in Johannesburg in 2014, I learned that under Mayor Parks Tau, the city began attacking this problem through a project called Corridors of Freedom, which anchors future development around new public transit arteries, especially in poorer neighborhoods outside of the city center. Easier and faster commutes will expand opportunities while also making neighborhoods along bus routes more attractive places to live, encouraging people to invest in homes and businesses. And by providing an affordable alternative to cars, it will also help reduce pollution and greenhouse gases. So the city's transportation strategy is also a climate strategy and an antipoverty strategy, all rolled into one.

Johannesburg is not alone in taking this three-in-one approach. The city of Medellín, Colombia, built an electric tram system (Metrocable) in 2004 to climb up the side of hills surrounding the city. It connects poor communities on Medellín's periphery to jobs and opportunities in the city center, while adding virtually nothing to the city's carbon footprint. New transit hubs created by the tram system have also become economic hubs, attracting businesses catering to commuters. When I attended the UN's World Urban Forum in 2014, I had a chance to see the system firsthand on a tour with the city's mayor, Aníbal Gaviria Correa. The tour provided a beautiful view of the city, and a great chance to practice my Spanish. My Spanish tutor is from Colombia, and I've told him I'm determined to become fluent before I die. At the rate I'm going, I'll be alive for a very long time.

Medellín is building on a great tradition of transit innovation in hilly cities. In the nineteenth century, San Francisco led the way with its cable cars, which back then were not a tourist attraction but a practical way of getting around. What will be next?

## THE DRIVERLESS SEAT

The possibilities for ride sharing are expanding exponentially with the advent of a new and very promising technology: autonomous vehicles. Autonomous cars have generated a lot of excitement because of their promise for reducing crashes and saving lives. But they can also offer other benefits for cities. They can create new transportation options that connect people to jobs and opportunities, breaking down economic barriers in communities and fighting poverty. And they could help speed the transition to EVs because owners of autonomous car fleets will find EV fuel savings highly appealing.

In addition, if autonomous cars lead to more ride sharing, individual car ownership may decline. In theory, that can help reduce traffic and the need for parking spaces and lots, leaving more room for pedestrians, cyclists, and businesses. It could also free up real estate currently used as parking lots. And if people spend less time looking for parking, pollution will decline and productivity will increase. In theory.

In reality, as the first waves of driverless cars hit our streets, it's vitally important we remember the lesson that cities learned the hard way over a century of car-based urban planning: We have to make sure that the technology serves the people of our cities, and not the other way around. And we have to make sure that it helps address existing problems and doesn't create new ones. After all, if people can spend time in the car doing things other than driving, they aren't going to mind being stuck in traffic, and this may lead to more cars on the road, more congestion, more sprawl.

It's true that autonomous cars could be the most important safety advance since the seat belt. But remember: the U.S. federal government didn't require seat belts to be installed in all cars until 1968. And we didn't start requiring people to actually use them until 1984, when New York passed the nation's first seat belt law. Think of how many lives could have been saved if we'd acted sooner. We can't make a similar mistake with autonomous cars.

Another concern: driving is a major industry that employs millions of people. Those jobs won't disappear overnight, but mayors have to plan for a future when the jobs may decline. Whenever there is disruption in an industry, labor markets are affected. Government has an important role to play in trying to help connect people in the middle of those disruptions to new skills and opportunities.

Cities are already rolling out pilot programs to introduce driverless cars. Car companies and tech firms are eager to be a part of that work, because it is a major new market for them. And since the challenges and opportunities that autonomous vehicles present are local in nature, mayors should take the lead on this issue, just as they have successfully on so many others. Mayors know the challenges and opportunities their cities face, and they are best positioned to explain them to the private sector, to form partnerships that improve people's lives, and to build public support for policies that put people first.

At the 2016 CityLab conference in Miami, I announced a new initiative funded by Bloomberg Philanthropies to help cities prepare for driverless cars. We're bringing together a group of innovative mayors who will create a set of recommendations for autonomous car policies, which can help inform cities around the world.

Not every transit idea is going to succeed. That's not how innovation works. But the ones that do are spreading from city to city faster than ever before, because of a number of intertwining trends. First, city leaders are talking with each other more and collaborating more than ever before, through organizations like C40 and the Global Covenant of Mayors for Climate and Energy. Second, it's easier to collect and report data, and more and more cities are doing it. This gives cities evidence of what works and confidence to invest precious resources in what will be effective—not just in transportation but elsewhere. And third, cities frustrated by the slow pace of national action on climate change are taking matters into their own hands.

By all indications, we're just at the beginning of a revolution in city transportation that is fundamentally changing the way we get around. As cities continue growing, so will the demand for transportation. But if we keep meeting that demand with twentieth-century automotive solutions, then by 2050 the average urban resident will spend more of his or her day stuck in traffic jams, with a life expectancy shortened by air pollution, in a city whose economy is slowed by car congestion, and in a climate characterized by unpredictability and extreme weather. That's not an appealing future—and fortunately, it's not inevitable. Through the choices we make, and the people we elect, we will determine our own fate.

# 10

# OIL'S TWILIGHT

*Oil dependence is a problem we need no longer have—and it's cheaper not to. U.S. oil dependence can be eliminated by proven and attractive technologies that create wealth, enhance choice, and strengthen common security.*
—Amory Lovins, Head of the Rocky Mountain Institute

In the spring of 1970, at the age of twenty-five, I returned from my work with the Peace Corps in India and became a newly fledged environmental lobbyist; almost immediately, I was assigned to work on the Clean Air Act. Getting ready required reading only one book, *Vanishing Air,* by one of Ralph Nader's Raiders. The first Earth Day had just occurred, and environmental advocacy was still in its infancy.

Ten months earlier, the California state assembly had rejected by one vote a bill to ban the sale of cars with internal combustion engines by 1975, which would have been a truly revolutionary change in the future of the automobile. The auto manufacturers prevailed, arguing that they could not sell cars in California if they could not sell IC engines. Later they conceded that if the bill had passed, they would somehow have complied.

Our job was to make sure that similar arguments did not derail the thrust toward a major national cleanup of air pollution. A lead Senate sponsor of the Clean Air Act, Maine Democrat Ed Muskie, was backstopped by Delaware Republican Caleb Boggs. Muskie wanted to be president, so he was bold. President Nixon didn't want Muskie to own the new, bipartisan anti-pollution issue—so the auto industry could not rely on the White House to weaken the bill.

Muskie put forth a proposal forcing the auto industry to devise new cleanup technologies and embed them in every car—a watered-down version of the California bill. The Big Three Detroit automakers sent their CEOs to Washington to explain to Muskie what could and could not be done. It backfired. Although Muskie made concessions to chemical plants and utilities through private lobbying, when confronted publicly by the auto industry, he hung tough.

I did not realize it, but California's failure to ban internal combustion, combined with Muskie's success in forcing Detroit to find and adopt new technology, would frame forty-five years of political combat over climate progress. Today, we are still fighting over the future of the combustion engine in transportation.

From 1970 until 2007, equipped with catalytic converters and other technology, internal combustion–powered cars and trucks got cleaner and cleaner. Engines also got more efficient. Instead of using this progress to reduce the amount of gasoline cars required (and wasted) however, the industry used this performance bonus to boost the acceleration (muscle) of ever-heavier cars.

California made another effort at moving beyond petroleum-powered transportation in the 1990s, establishing a mandate that first 2 percent, then 5 percent, and eventually 10 percent of car sales be "zero-emission vehicles" (ZEVs). The pushback was immediate and overwhelming. Resistance from auto companies and the oil industry, combined with legal victories by oil concerns, forced California to abandon the mandate and GM—controversially—to recall and scrap the entire fleet of its first-generation electric car, the Impact.

Over this forty-five-year period, cheaper or better fuels took oil's place in generating electricity, making chemicals and plastics, and heating buildings. And yet, oil retained its monopoly as a transportation fuel. Even in 2015, 92 percent of total U.S. transportation fuel came from oil. Why? Because other fuels—electricity, natural gas, biofuels—don't have meaningful market share. This monopoly positions oil for most Americans as a "necessary evil," enabling the industry to get away with price-gouging, environmental destruction, and climate disruption because there is no alternative.

In the spring of 2016, in both the United States and Europe, oil passed coal as the biggest source of climate risk. Globally, oil now accounts for 34–36 percent of fossil fuel emissions. Numerous analyses point out that finding a substitute for oil in transportation is essential to solving the climate crisis. But—just as it did in California in 1969—the oil industry continues to claim today that no other fuel can power vehicles. And they have worked hard to make sure this remains true.

The American public, according to research, would like to break with our dependence on oil—they think that it is dirty and expensive, and that it impairs their health and our nation's security. They also believe that it is technically possible to replace most oil with better fuels, and they are correct. Yet, while 70 percent of Americans think that the United States should replace oil as a transportation fuel over the next half century, only 50 percent think we will manage to do it, because of the political power of the oil industry.

I'm in the other 50 percent, because the building blocks to do it are already available to us. Let's begin with a simple step: Use as little oil as we need to move every car and truck.

## HIGH-PERFORMANCE VEHICLES

The Sierra Club long campaigned to improve the efficiency of the combustion engine. We focused on the idea that "The Biggest Single Step" the United States could take to protect the climate was to require Detroit to improve fuel economy in its cars. The first fuel economy (CAFE) standards dated to the 1970s, when gas station lines drove Congress to act, but they weren't updated for decades. Even worse, Detroit had outrageously exploited a loophole that allowed larger passenger vehicles to be built on pickup truck chassis and thus exempted from the requirements for sedans. (This was the birth of the SUV boom. Sloppy policy has consequences.)

With oil cheap and seemingly abundant during those years, the auto industry was back in the driver's seat.

Dan Becker, the club's auto lobbyist, and I made regular pilgrimages to Detroit to meet with the auto companies and unions. At one 1990s meeting with Steve Yokich, then United Auto Workers president, I urged him to look ahead. Detroit's whole business plan depended on building SUVs on outmoded pickup truck technology and marking them up enormously. I asked Yokich what the industry would do when Japanese and German makers began building big cars with modern technology. He took me to the window of Solidarity House, UAW headquarters, and showed me the parking lot. "What do you notice?" Yokich asked me. "Well," I said, "they're all American cars." (Out my window in San Francisco, they were not.) "True," he said, "but also, almost no SUVs. Our members know crap when they build it." He was right. Even so, when I asked him to join us in forcing the American auto industry to start innovating, he referred me back to the car companies. And when the Sierra Club proposed a major tax credit for hybrid vehicle production, Detroit's lobbyists declined the offer.

George W. Bush's defeat of Al Gore in 2000 strangled the tax credit's chances. It also stripped the United States of the budget surpluses that made the credit affordable in the first place. But then in 2004 the price of oil began to soar—from $30 a barrel up to an eventual $110—while U.S. auto companies continued to bet their business on the fantasy that gasoline would never cost more than $2.50 a gallon. It was a very bad bet. As

China drove global demand for oil up ever higher, U.S. gasoline prices crossed the $2.50 mark in September 2006, topping $4 before the sub-prime crisis crashed the global economy. GM and Chrysler faced bankruptcy; Ford survived behind huge lines of credit but it was hurting. Detroit's resistance to innovation now threatened not only its profits but its very existence.

Faced with the collapse of the domestic auto industry, the Bush and then the Obama administrations bailed out the industry. But Obama required the new leadership of GM and Chrysler to embrace innovation. The administration also adopted much tougher fuel economy standards, rules that will cut gas consumption by 50 percent by 2025, saving U.S. consumers $1.7 trillion at the gas pump, reducing U.S. oil consumption by 12 billion barrels, and cutting atmospheric $CO_2$ by 4 billion tons. At the same time, the UAW, under a new president, Bob King, strongly supported the efforts by the Obama administration to force U.S. automakers to invest more in advanced, clean-vehicle innovation.

As a result of these changes, and buttressed by continued high oil prices, fuel economy for cars increased from 30 to 35.6 miles per gallon by 2014. This is a significant improvement, but much more needs to be done. In order to cut U.S. oil consumption by 50–75 percent—the amounts needed to protect the climate and protect against future oil price spikes—the internal combustion engine must be replaced by zero-emission vehicles powered either by batteries or fuel cells.

## EUROPE'S DIESEL ROMANCE GONE SOUR

Europe took a different path. Because the continent had to import its oil, it used high gas taxes to drive consumers to smaller vehicles. And as climate became an important issue, Europe doubled down on diesel, which is generally more efficient than gas. Diesel pollution was dismissed as a mere engineering barrier, to be solved by improving pollution control.

In 2003, before oil prices began to soar, Toyota released the redesigned version of its hybrid Prius into the U.S. market. Volkswagen, lusting to supplant Toyota and GM as the world's biggest auto producer, countered with a particularly large bet—not just on diesels, but on smaller diesels, in spite of the fact that it was difficult to fit diesel pollution control equipment into such small cars.

When VW's small diesel engineers could not meet stringent U.S. clean air emission standards, particularly on oxides of nitrogen, they tried again—and again. Eventually, after repeated failures, VW resorted to fraud: It installed a "cheat" device that turned pollution control systems on whenever the car was being tested, and off whenever it was on the road. On the road, in fact, VW's new "clean diesels" emitted up to forty times as much nitrogen oxide (NOx) as allowed. Meanwhile the company's ads trumpeted an engineering triumph.

For five years the fraud went undetected, but in 2015, independent reviewers found the cars were massive polluters. The rest is probably familiar: VW revealed the fraud and its market capitalization fell by a third. Sales plummeted. An initial settlement with the federal government is expected to cost $15 billion ($30,000 per car).

Alerted by VW, auto regulators everywhere began scrutinizing other manufacturers. GM's European division, Opel, was also found to have installed "cheat" devices. Larger Mercedes-Benz diesels allegedly turned off their NOx pollution control systems—albeit only in cold weather. Fiat programmed its diesels to turn off emission controls after twenty-two minutes of driving. Mitsubishi lost half of its market value—and ended up being acquired by Nissan—because for ten years it misrepresented the fuel economy of its fleet.

By mid-2016 it was clear that almost all of the world's diesel passenger fleet was violating pollution control limits under actual highway conditions, at least in some seasons. But, for the moment, and backed in many cases by their home governments, most manufacturers had post-poned abandoning diesel cars, even ones that didn't meet standards.

This revelation explained why so many European cities had been struggling with elevated pollution levels even after the first of nominally tough pollution rules. When the deceptions were revealed, communities fought back by proposing to ban diesels altogether.

The European public, as reflected in the EU parliament, want strong action to move away from oil-fired vehicles and their endless pollution threats. But European governments are divided. Some want to move on to electrification. Many fear that they will lose market share to Japanese and American companies that bet on electrification earlier. How long Europe will cling to its diesel romance remains to be seen, but the market is clearly moving against diesel cars and towards electrification.

While diesel engines may be replaced for cars, there are many heavy-duty uses (trucks, construction equipment, ships, and trains) for which they are likely to remain a workhorse for a long time to come. Diesels emit soot particles—black carbon—particularly if they are powered by fuels that contain significant sulfur. This is a very serious problem: About 25 percent of global black carbon comes from either stationary or mobile diesel engines. Luckily, this is one climate problem with a simple technology fix. Black carbon from diesels can be nearly entirely eliminated by using ultra-low-sulfur fuel and diesel particulate traps, which are now standard equipment on both large and small diesel vehicles in the United States. In emerging economies, the right financing tools can accelerate the adoption of black carbon control measures—just as they can with solar panels.

## FREIGHT: THE LITTLE ENGINE THAT COULD

One of the biggest barriers to reducing climate pollution almost never gets mentioned. It's getting goods—wheat, shoes, two-by-fours, tables, refrigerators, iPads, coal, oil, and solar panels—from the mine or factory to the store. Shipping goods creates about half of today's transportation emissions, and this volume is projected, unless we get smarter, to increase by 45 percent by 2040.

Here, uniquely, the world doesn't need new solutions. The two oldest modes of transporting goods—ships and trains—are by far the least polluting. Ships are the most efficient way to move goods. Even inland barges carry 600 tons of goods a mile for every gallon of fuel they burn, about three and a half times as efficiently as trucks. Unfortunately, waterways don't go everywhere. So efficient movement of stuff in most places means trains.

The United States has one of the best freight rail systems in the world. It moves a ton of freight almost as efficiently as most ships, carrying 40 percent of U.S. cargo (trucks carry 33 percent). Globally, trains carry twice as much freight per gallon of fuel used as trucks.

At present there are wide variances in how much countries take advantage of rail's essential role in saving fuel and protecting the climate. Admittedly, many are limited by geography, or financing, or both. But there is certainly more that most nations can do to become rail-intensive, and hence climate-friendly.

The EU uses rail for only 18 percent of its goods, and allows restrictive and monopoly practices that discourage growth. India used to ship about 70 percent of its goods from the railroads left behind by the British; now that has fallen to a third. Latin America is woefully under-railed, with 1 percent market share of goods while trucks get 30 percent. Africa's few railroads were built to run from old mines and plantations to the coast, rather than to connect cities and countries. Heavy commodities like iron ore and fertilizer are carried long distances by truck in Africa, raising prices enormously and holding back prosperity.

So where are development agencies and banks investing? In road, not rail. Asian Development Bank loans for transportation run 74 percent to 15 percent road to rail; World Bank funds go 60 percent to road, 40 percent for all other forms of transportation combined.

# Sources of Climate Change

## By Major Pollutants

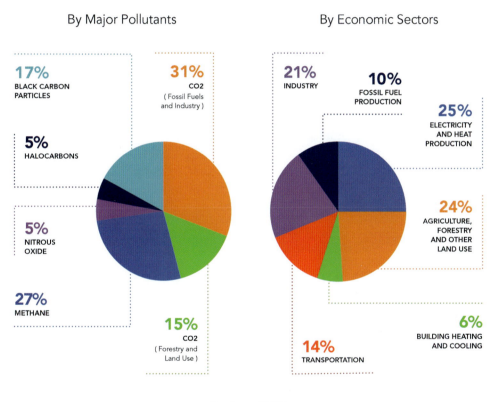

**17%**
BLACK CARBON
PARTICLES

**31%**
CO2
( Fossil Fuels
and Industry )

**5%**
HALOCARBONS

**5%**
NITROUS
OXIDE

**27%**
METHANE

**15%**
CO2
( Forestry and
Land Use )

## By Economic Sectors

**21%**
INDUSTRY

**10%**
FOSSIL FUEL
PRODUCTION

**25%**
ELECTRICITY
AND HEAT
PRODUCTION

**24%**
AGRICULTURE,
FORESTRY
AND OTHER
LAND USE

**6%**
BUILDING HEATING
AND COOLING

**14%**
TRANSPORTATION

Data Source: US EPA

# Is Climate Changing? Ocean Heating Tells the Story

Air and Water Temperatures on December 25, 2015 Against Historic Averages

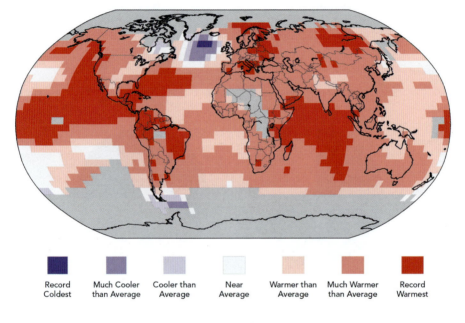

Record Coldest | Much Cooler than Average | Cooler than Average | Near Average | Warmer than Average | Much Warmer than Average | Record Warmest

Data Source: NOAA's National Centers for Environmental Information

## Coal's Declining Market Share

Share of Coal Generated US Electricity: ● Coal Share %

## Wind and Solar's Price Decline

Cost of Electricity Generated By: ● Wind ● Solar

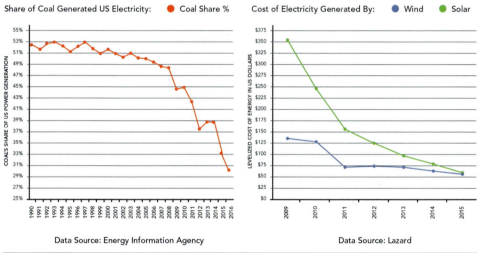

Data Source: Energy Information Agency

Data Source: Lazard

# We Can Learn from Old Cities

Dark Rooftops in Modern Chandigarh

White Rooftops in Historic Jaisalmer

Data Source: Google Earth

# Soil Degradation

Deforestation and erosion release carbon.
Restoration and reforestation can pull it back out of the atmosphere.

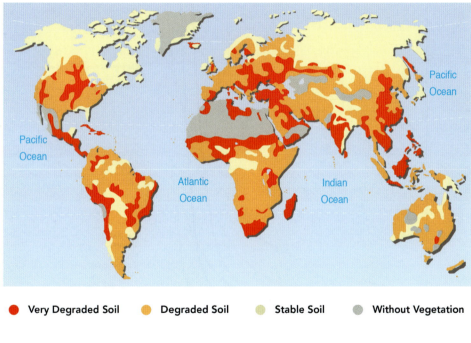

Very Degraded Soil    Degraded Soil    Stable Soil    Without Vegetation

Data Source: UNEP, International Soil Reference and Information Centre (ISRIC), World Atlas of Desertification, 1997.
Philippe Rekacewicz, UNEP/GRID-Arendal

# The Transportation Revolution is Here

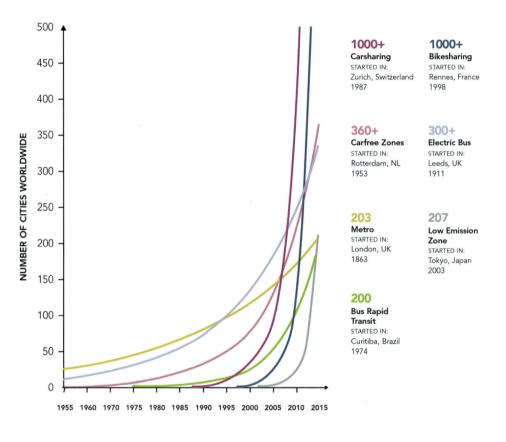

NUMBER OF CITIES WORLDWIDE

500
450
400
350
300
250
200
150
100
50
0

1955 1960 1970 1975 1980 1985 1990 1995 2000 2005 2010 2015

**1000+**
**Carsharing**
STARTED IN:
Zurich, Switzerland
1987

**1000+**
**Bikesharing**
STARTED IN:
Rennes, France
1998

**360+**
**Carfree Zones**
STARTED IN:
Rotterdam, NL
1953

**300+**
**Electric Bus**
STARTED IN:
Leeds, UK
1911

**203**
**Metro**
STARTED IN:
London, UK
1863

**207**
**Low Emission
Zone**
STARTED IN:
Tokyo, Japan
2003

**200**
**Bus Rapid
Transit**
STARTED IN:
Curitiba, Brazil
1974

Data Source: World Resources Institute

# Chinese Pollution Crosses the Pacific

Pollution from China harms East Asia and the Western United States.

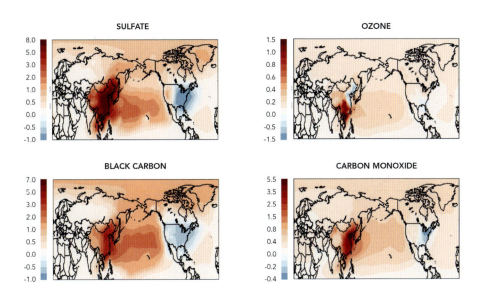

Data Source: Proceedings of the National Academy of Sciences

## Projected Decreases in California's Snowpack

The mountain snowpack in the Sierra Nevada provides water
to the majority of Californians. By the end of this century, as
little as 20 percent of the Sierra snowpack may exist.

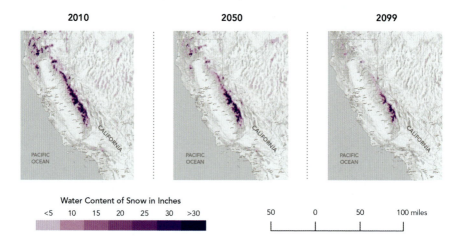

Data Source: Data by OpenStreetMap; Map Tiles by CARTO, Stamen Design, ASTER GDEM

What's the source of this bias? Roads are easier to build because they have more local benefits, require less complex planning, and often lend themselves to more corrupt political bargaining. Rail, on the other hand, requires maintenance, something governments in developing countries often overlook. African nations can't agree on how to connect to each other by rail. And when Brazil's president, Dilma Rousseff, embarked on a huge rail investment, it was financed by—and designed to benefit—China, which wanted cheap access to Brazil's commodities. A better idea would have been to connect Brazil's internal economy more closely and productively.

Getting rail right is not easy. But rail is one of the most powerful economic development tools ever created. We know how to do it, and it would drive down climate pollution faster than almost any other simple transportation trick we could adopt.

Even if we get rail right, however, much of the world's freight will continue to move, particularly for the last few miles, by truck. More efficient trucks slash emissions and save money. Here the world is moving. The Obama administration enacted its second round of fuel efficiency standards for trucks, with support from shipping companies that want to save money. Europe is following suit, after slapping truck manufacturers with a fine of 3 billion euros for conspiring to slow progress on fuel savings. India is close behind. So common sense is yielding progress. Better late than never.

## TAKING FLIGHT

Aviation accounts for only 3 percent of climate emissions, but it's the fastest-growing transportation sector, projected to increase three to four times by 2040. The global community took a significant step forward in 2015 when almost all of the major aviation countries agreed to a new approach in which total aircraft emissions would be capped at 2020 levels, and any increase after that date that could not be avoided with more efficient aircraft and operating rules would have to be offset by paying to reduce emissions in some other sector.

This "market-based mechanism" is a good start, but, over time, the offsets are likely to become a problematic loophole. And if the global economy is strong, even a carbon tax won't cut demand for air travel. The best way to lower airplane emissions is by redesigning aircraft, and the industry itself, to be more efficient.

Airlines vary widely. The least efficient use three times as much fuel per mile as others—emitting three times as much climate pollution. Some of this difference results from the age and efficiency of the fleet, or a carrier's ratio of short-haul to long-haul routes. But most is due to operational issues, like optimizing the amount of freight and number of passengers on each flight. Most important is how much travel is nonstop. A quarter of airline carbon emissions occur during takeoff and landing, so direct flights cut emissions dramatically. Efficient routing and operations save airlines and their passengers fuel, money, and time. Why, then, doesn't it happen? Why aren't all carriers as efficient as the best? Well, one reason is that the hub-and-spoke system allows certain airlines to effectively monopolize hub cities. These airlines fiercely resist proposals to let their competitors fly more efficient direct routes. Other steps to improve the industry's efficiency, including improving air traffic control systems and incentivizing larger planes, would also help reduce emissions.

Airline travel could be much less climate-threatening—but we would have to make it much more efficient and competitive to get there.

## REINVENTING THE CAR

On a warm September day in 2010, I stood outside the U.S. Capitol to watch the X Prize Foundation announce the winner of its $5 million competition for designing a car that gets 100 or more miles per gallon. House Speaker Nancy Pelosi introduced the winning firm's founder and CEO, Oliver Kuttner, a racing car devotee and real estate developer with a broad smile. His firm, Edison2, was based in Lynchburg, Virginia, and I was struck when he broke from his prepared remarks to explain why. In short: the town is a former Atomic Energy Commission site that boasts a wealth of small firms with expertise in material science and fabrication—machine shops that can help solve the thousands of problems that arise from designing a car that, with a steel body and an internal combustion engine, still gets more than 100 mpg. "We're in danger of losing those kind of skills here in the United States," Kuttner said. Captured by his vision, I introduced myself to him. He represented one end of the innovation spectrum that I'm confident will crack this dilemma—I just don't know when. His goal is to make cars so simple, so light and cheap to build, that we can slash their need for fuel—whatever the fuel—to a fraction of its present level.

At the heart of Kuttner's disruption lies a problem I never knew about: the suspension that today's cars borrowed, relatively unchanged, from horse-drawn coaches. The standard suspension rests above the axle. Each wheel connects through three stress points—each one of which, in turn, must be rigidly connected to the other eleven. Kuttner put the suspension in the wheel hub, enabling a single connection for each wheel, four for the whole car. A lighter, simpler chassis meant a much lighter car, with fewer parts.

Kuttner's 2013 prototype, electrically powered, was rated at 245 mpg. It weighed one-third of its conventional counterpart, using aluminum instead of steel, and had cut the part count by two-thirds. But Kuttner had still not broken through in the world of automotive manufacturing. The engineers who worked for the auto companies were unwilling to embrace an outsider's vision. The existing venture capital models for disruptive innovation rejected automotive start-ups as too slow and expensive. Between the fall of 2010 and 2015, most of the truly innovative companies attempting to move from prototype to mass production in the automotive space had failed: Fisker, Bright Automotive, Next.

The one exception, of course, is Tesla. Tesla makes its cars in a Fremont, California, plant with which I had a long history. In 1984 I agreed, pro bono, to advise NUMMI, the GM-Toyota joint venture that was taking the plant over from GM. When the NUMMI team inspecting the plant discovered a huge old crack in the paint bay, through which millions of gallons of paints and solvents had contaminated the soil underneath for decades, the GM response was: "Call the lawyers." Toyota overrode them: "Call the EPA and do whatever they ask." That became the new template. Within a few years, the plant, which GM had used to produce North America's shoddiest cars, was producing the best cars in the United States. When the recession caused NUMMI to dump the plant, Tesla snapped it up.

In 2015, I took my first tour of Tesla's plant. If NUMMI was light-years more innovative than the old GM plant, what I see today is more revolutionary still. The plant was quieter and cleaner than many open-plan offices. The heavy work was done by robots.

Tesla CEO Elon Musk outlined the ideas behind this transformation. The internal combustion engine, he argued, "is an enormous kludge—all those cylinders and millions of explosions to manage." It's remarkable, he pointed out, that engineers can make these engines reliable—which, in his (and my) youth, they were not. "You didn't take it for granted that the car would start," he recalled.

Instead, Musk had a powerful vision: True innovation and sustainability demanded that he design from first principles—from the ground up. That's why Tesla customers rave more about its performance than its fuel savings. "The engine block of an internal combustion engine is a very heavy mass, perched high, up front, on a rubber mount," Musk explained. "It's like trying to steer a big bobble head around a curve. We get to put the battery, the center of gravity, low and in the center. It's a totally different experience." It is. I've driven one.

Challenged on the $5 billion investment that Tesla was about to make in its Gigafactory to produce lithium-ion batteries, Musk pointed out that it would take 200 such factories to turn out the 100 million EVs needed to replace the gasoline engine. "There's no choice but to go to scale," he said.

Musk is not alone. Toyota has promised to sell only hybrid and hydrogen vehicles by 2050. China's biggest car company has invested $3 billion in EVs. Auto shows are now dominated by electric or fuel cell vehicles. Oil's monopoly is being threatened.

## TOWARD A NEW MOBILITY REVOLUTION

Autos evolve slowly, because it's very hard for newcomers to get good at building them—even Elon Musk has a hard time with the manufacturing side. But automobile companies have steadily innovated in one space: Cars have been digitized. With each succeeding model year they become ever more computerlike. Indeed, Ford's ultra-expensive new GT sports car employs twenty-eight microprocessors and more lines of code than a Boeing 787 Dreamliner. While today's automobile may look and handle much like its 1960s predecessor, the crucial components are no longer steel and rubber but silicon and megabytes.

That's significant for a key reason. Modern airplanes are flown not directly by their pilots but by instructions their pilots feed into the plane's computer system. The computer then instructs the engines, wings, flaps, and rudders. And most of the time, those orders have been preprogrammed into the autopilot function—if you travel overseas, most of your annual travel miles are already computer-controlled, in a vehicle produced by Boeing or Airbus. A similar evolution began for automobiles in 1965 with cruise control. And as car manufacturers steadily enhanced the sophistication of cruise control, a vision emerged of cars that would drive on autopilot, like a 787, or eventually dispense with the need for a driver altogether—becoming a so-called fully autonomous vehicle.

So just as Tesla was demonstrating the superior performance of electric vehicles, and Edison2 was forging new pathways to simple, cheap, and hyper-efficient chassis and bodies, the auto and tech industries were getting ready to take the driver out of the equation. Elon Musk predicted that half of all passenger trips will be autonomously controlled by 2030. And Silicon Valley companies like Apple and Google are trying to compete with the auto industry in leading the autonomy revolution. Meanwhile, a totally different kind of innovation was getting set to further disrupt the way we get around. Instead of making cars simpler or changing their power source, Silicon Valley innovators decided to change how they are owned.

The San Francisco headquarters of Lyft sit in the Mission District, one of the major start-up neighborhoods in the city. Upon entering, you sign a nondisclosure form. Inside, I met with John Zimmer, one of the two founders of the company. The idea that became Lyft popped up in Zimmer's first year at Cornell University's famed hotel school. The basic course in hotel management, Zimmer explained to me over coffee, came down to this: "There are only two things that count in hotel management. The customer experience and occupancy factor. Nothing else matters." In the auto industry, the average American spent (at that time) $9,000 owning and operating a car that then sat idle 95 percent of the year. "That," said John to himself, "is a pathetic occupancy factor. And the customer experience? You buy the car at a showroom you hate, clean it yourself, fix flat tires, and if it breaks, have to get it to the shop on your own. There has to be something better." That something better became Lyft, the idea that people could share their own cars.

Suddenly partnerships with California tech firms became attractive to the big automakers. Ford opened a "future of mobility" research facility not in Michigan but in Silicon Valley—the better to keep an eye on the new competition. In early January GM announced a partnership with Lyft to work together for the development of autonomous vehicles, only weeks after Ford signed up with Google. Disruption is in the air—and none of it depends on oil.

The automotive age that began with the Model T rested on three legs: Cars were gasoline-powered, driver-operated, and privately owned. They were carriages powered by gasoline. But in the last three years all of these legs have been challenged—the power source by Elon Musk, the need for the driver by the auto companies, and the ownership model by Lyft, Uber, and others. This is the "triple mobility revolution" now under way.

Governments are going to play a key role. During the Paris summit, four European nations, including Germany and the United Kingdom, joined nine U.S. and Canadian states and provinces, including California and New York, in pledging an end to gasoline- and diesel-powered cars by 2050. The German Bundesrat passed a resolution in 2016 calling for an end to combustion vehicles by 2030. While predicting innovation is a fool's errand, there are good reasons for suspecting that the combination of autonomous/shared/electric vehicles is a potent one that will redefine our experience of the road in the near future. The three strands reinforce each other. Driverless cars make an Uber or taxi fleet much cheaper. Such fleet vehicles are driven many times farther each year than personal cars, making the fuel savings of an electric fleet very attractive. And electric cars are easier to design to be driverless. A green overhaul of transportation promises to be the real clean-tech revolution—more significant even than the expansion of wind and solar energy.

Who will lead this revolution? The American edge cannot be taken for granted. The Obama administration was highly committed to clean energy, but it missed an obvious opportunity when it allowed the U.S. Postal Service to duck the chance to use its mass purchasing power to convert to electric delivery trucks. It's not clear what the posture of the Trump administration toward transportation innovation will be, but cities and states can lead the way—just as China is.

The Chinese government set a target to put five million EVs on the road by 2020. Chinese investor Sunny Wu is one of the middlemen accelerating the transfer of clean energy technology from the United States to China—often before it even gets to market in North America. *Fortune* outlined his strategy as follows: "Buy Western companies that have good technologies but poor domestic growth prospects and bring them to China, where Wu and his contacts serve up the money and the market to help the firms grow very big, very fast."

In January 2012, I was astonished to hear what Wen Jiabao, the Chinese premier, said to a Clean Energy Summit hosted by the world's ninth-largest oil power, the United Arab Emirates. Standing in the heart of the global oil industry, Premier Wen called for a revolutionary reform of oil markets, arguing that they are driven by cartels and speculators, with prices unrelated to supply-and-demand fundamentals, and intolerably volatile. He suggested a partnership among the G20 countries, mostly oil importers like China and the United States. His Persian Gulf hosts took notice and placated China with some favorably priced oil sales. But China's potential partners, the United States and the EU, ignored him and his offer to collaborate to end oil profiteering by OPEC. Nevertheless, the opportunity is still there.

Seizing it, and working together to break oil's monopoly in transportation, would also position the U.S. to lead the looming transportation revolution. But if global competitors outpace the United States in developing EV markets and reducing dependence on oil, the 2009 rescue of General Motors and Chrysler may turn out to have been a brief respite for the U.S. auto industry. As we have with wind and solar, we risk ceding our early advantage in transportation to competitors overseas. The rewards of innovation are not captured by those who merely invent it. The value-added supply chain does not settle where a technology originates, but where the market for it scales.

If the United States wants high wages and economic returns from the emerging world of autonomous transportation, state and federal governments must go beyond promoting early innovation—they must help drive early adoption. Robust markets are what enable the social benefits of new technologies. Societies that don't adapt to that reality risk getting left behind.

## OVER A BARREL

Understanding the market for oil is critical to solving our dependence on it, which is one of the biggest threats to the climate. Most of the world's remaining easy, cheap-to-pump crude lies in the Middle East, Russia, and Venezuela, managed by government oil companies. These government oil companies—sometimes to create artificially high prices (OPEC), sometimes because of political mismanagement (Venezuela)—don't pump nearly as much oil as their reserves would allow. Private oil companies—Exxon, BP, Shell—hold reserves of low-cost "legacy" oil that they discovered years ago. But to meet peak global demand, they have been forced to develop expensive, hard-to-get-at oil that can cost more to extract than the market will pay for it.

Private oil companies do invest a lot in new fields when prices are high, but it then takes five to ten years to bring these new discoveries to market. And oil consumers buy more efficient vehicles when oil is expensive, big SUVs when it is cheap. This combination—production constraints on low-cost oil by national oil companies, long lead times for new oil fields developed by private companies, and a consumer tendency to respond only to the short-term, immediate high or low price of oil—creates a very volatile market. Oil prices are a roller coaster, with periods of peak supply and low demand created by high prices, followed by supply shortages and demand spikes.

During a price boom, countries sitting atop low-cost reserves get to charge the highest price required by the most expensive barrel. These unearned profits are termed "petroleum rents." From 2006 to 2014 these rents resulted in a massive transfer of wealth from oil-importing nations— the United States, the EU, Japan, China, India—to exporters like OPEC and Russia. At one point, five cents of every dollar of global GDP was being paid as a windfall profit to oil-exporting countries, crippling growth in places like India and Africa, helping to bankrupt GM and Chrysler, funding extremist politics in Venezuela and Saudi Arabia, and costing the U.S. economy trillions of dollars. But those high prices encouraged the boom in U.S. shale oil, and led drivers to seek out more efficient vehicles and governments to set tougher efficiency standards for cars and trucks. Growth slowed in India and China.

Then the cycle flipped. The shortage of oil on the market became a glut—not a big glut, just a few percent. But that drove oil prices down from $110 to below $50 a barrel. Saudi Arabia and Iran were still making money, but most of Exxon's and Shell's recent discoveries were not. The total savings from the shift in the price of oil have been $1.5 trillion each year. Now that oil is down in the $40–$60 range, there is a real chance for the largest consumers, like the United States, Europe, Japan, India, and China, to keep it there, and a $1.5 trillion incentive to do so.

Some economists point out that a carbon or oil tax would discourage oil consumption and cut prices. That's true, but it wouldn't be enough. The soaring price of oil from 2004 to 2014 was equivalent to a $200 tax on carbon. Despite that punishing reality, consumption didn't fall. High prices and taxes don't work very well as disincentives when the customer has no choice but to buy the product anyway. Oil's 90 percent dominance of transportation energy is not because it is cheap, but because it has no competition.

Even though natural gas trucks and electric cars are cheaper to operate than diesel or gasoline, consumers still have no effective choice—the oil industry controls the fuel distribution system and prevents competition. Manufacturers of EVs have trouble selling them because customers fear that they won't be able to charge up. New technologies often have this problem. They can't cross over what Silicon Valley calls "the valley of death," the place where the technology is ready before the market is. But we know how to encourage competition using government incentives, as we did over the past fifteen years for renewable electricity. To reduce the cost of a new technology, enlarge the market capable of buying it.

Right now, electric cars don't enjoy the economies of scale that they need to become cheaper. Different countries (and states) are experimenting with a series of incentives, such as low-carbon fuel standards. A number of U.S. states have EV sales mandates, and the EU is considering joining them. Norway provides generous tax relief, Germany helps offset the cost of electric cars, Chinese cities make it easier for owners of EVs to license their vehicles, and, as Mike mentioned, Singapore is rolling out an electric car–sharing service.

Providing this kind of support for electric cars may seem expensive, but only when you consider it narrowly: For major oil-importing countries (the United States, China, India, Japan, S. Korea) and the EU, it functions as an economic boost. Increasing the market share of EVs, combined with extending and improving today's fuel economy standards for cars and trucks, will reduce demand for oil. Reduced oil demand will mean *dramatically* lower prices for the oil that we still need. Aggressive support for electrified, efficient cars and trucks could reduce global oil demand by eleven million barrels a day by 2030, cutting oil prices at least 40 percent. That $1.5 trillion annual savings in the cost of fuel will pay for EV subsidies many times over—and provide more savings to be used toward low-carbon infrastructure.

But if demand for oil is allowed to surge again, because we don't invest enough in new technologies like greater-efficiency or electric vehicles, another oil shock will surely once again send the economy into a tailspin.

The chance to lock in moderately priced oil and end the cycle of predatory petroleum rents is a stunning opportunity—probably the single biggest economic opportunity our climate problems offer. No other simple market failure costs the world as much as the failure to break oil's transportation monopoly and create competitive fuels for moving people and goods. Once there is competition, the global economy can transition to a future with clean, renewable transportation. And climate-destroying, remote, and costly frontier oil reserves can remain where they belong— in the ground.

# PART VI

# COOL CAPITALISM

## 11

# WHAT WE MAKE

*A sustainable world is possible if we take advantage of the*
*vast opportunities in manufacturing. We must view these*
*industries—and their supply chains—as a source of*
*solutions, not just a source of problems.*
—DIANE REGAS, EXECUTIVE DIRECTOR,
ENVIRONMENTAL DEFENSE FUND

At the end of summer 2016 Great Britain released some perplexing numbers on its climate emissions. Domestically, British emissions had fallen to a promising 26 percent below 1997 levels. Overall, though, Britain's climate footprint had grown by 3 percent. Why? Because footprint calculations include both domestic emissions and emissions associated with imports—food, timber, cars, clothing, construction steel, equipment, computers, toys. These "imported emissions" are now 55 percent of Great Britain's total climate impact, and they rose by 7 percent in 2016. Imported emissions lie at the heart of the challenge posed by manufacturing, the final major sector of the economy driving the climate crisis.

Manufacturing represents about one-fifth of our impact on the climate. A large portion of these emissions are created in one country and consumed in another—call it the climate pollution trade. Most countries that export carbon-intensive goods are emerging markets like China and Brazil. Most of the biggest emission-importing nations are advanced industrial nations like Britain, Japan, and the United States.

(The oil-exporting countries discussed in Chapter 10 are in a league of their own.)

Making stuff—smelting iron and steel, weaving textiles—was the bedrock of the Industrial Revolution and the beginning of fossil fuel dependence. Coal was the breakthrough drug in this dependence, because Britain, abundantly supplied with coal, was the first mover in mechanized production. Industry uses carbon fuels for three main purposes. First, as power—to run pumps, turn wheels, power motors. Second, for heat—to melt glass, dry grain, fire bricks. Third, as a chemical input—into processes like smelting iron or firing cement, in which the carbon in coal bonds with oxygen in iron ore or limestone. This leaves behind purified iron or cement stripped of its contaminating oxygen atoms but also creates massive quantities of carbon contaminated with that same oxygen, $CO_2$.

Reducing the emissions associated with manufacturing gets relatively little attention in the climate conversation, partly because so many of these emissions occur across an ocean or two. But it matters a lot—and it matters where things are made. Making aluminum in China takes seven times the carbon of smelting in the EU; Chinese steel is three times more carbon-intensive. Polypropylene mills in China gobble twenty times as much carbon as similar mills in Europe.

## GAS LEAKS

The biggest challenge we face when thinking about how to reduce manufacturing's climate impact is also the easiest to address: capturing methane. Reducing emissions from manufacturing starts with cleaning up the oil and gas industries that power the factories.

So much methane leaks from oil and gas wells or pipelines that by 2018 the oil and gas industry is projected to account for 90 percent of U.S. methane emissions. Remember, in the short term, methane is eighty-four times as good at holding on to solar heat as $CO_2$. (It also breaks down quickly, so once we get on top of methane emissions, we could make climate progress very fast.) The oil and gas industry's methane emissions come from two main sources: leaks at the wells that drill for oil and natural gas, and in the pipelines that deliver natural gas.

Checking and monitoring the pipes for leaks is relatively easy, but it is not cost-free, and the industry has been fanatical about cutting costs. So while better inspections would reduce leaks and thus increase revenue, companies may not consider it cost-effective. The same is true of leaky wells and storage facilities, even though such leaks can pose serious health and safety risks. In California in 2015, Aliso Canyon, an old oil field re-purposed as a natural gas storage facility, was found to have enormous leaks, venting about a thousand tons of methane a day for three months. This discovery forced not only shutdowns in the California natural gas supply chain but also the evacuation of entire neighborhoods. The leak was the result of weak regulation of the wellheads of such oil fields. A 2010 pipeline explosion in San Bruno, California, killed eight people, resulting in a $1.6 billion fine against PG&E, the utility operating the pipeline. In 2016, a Canton, Illinois, blast killed one and injured many more.

Pipeline companies lag badly in performing the required maintenance and inspection ostensibly required by federal law, and yet the government has routinely allowed them to postpone compliance with safety standards. The industry has then taken advantage of this regulatory vacuum to engage in a wide range of dangerous polluting practices.

Lack of enforcement is not the only problem. There is also a lack of smart regulation. As a result, oil companies have made little effort to capture the methane that escapes during drilling. Instead, they burn it on the spot—a practice called flaring. Drillers in North Dakota's Bakken oil fields flared so much methane that at night, from a satellite, the state shone as brightly as Chicago. And yet that was legal. In offshore oil fields in Nigeria and Angola, virtually all of the natural gas coming out of oil wells is flared, even though a few miles to the east are economies desperately short of electricity (which natural gas can generate) and fertilizer (for which natural gas is the major feedstock). This happens in spite of requirements in leases that natural gas be brought to market.

Corruption plays a role here. Building pipelines to get the gas to useful markets is expensive for the oil companies, so it is often easier for them to persuade local officials to ignore the requirements and let flaring continue—sometimes even bribing them to do so. The oil industry globally has fought vigorously to keep its leasing contracts with governments confidential, and it is this secrecy that makes such bribery possible. U.S. companies, at least, are forced to disclose their terms and payments, so there is some check on these kinds of corrupt practices—but not enough of one. (Unfortunately, in 2017, Congress voted to roll back the requirements that U.S. oil companies disclose their leases with foreign governments.) We need more rigorous monitoring and measuring of natural gas production and distribution around the world—not more loopholes.

In the United States, some local governments have passed their own ordinances to protect against wasteful and harmful practices, but the oil industry has also had success getting sympathetic state legislatures to strip those localities of their ability to do so. And it's possible that President Trump will reverse basic methane safety rules put in place by the Obama administration. But, as we have seen through the Beyond Coal campaign, citizens and communities have power when they organize and fight back. If there are clear, enforced, and consistent rules for oil and gas producers, extracting and producing fossil fuels will be a lot less dangerous to the climate.

## THE TIMBER RACKET

While the oil and gas industries exploit their exemptions from environmental laws, another global industry that is essential to manufacturing—

timber—routinely ignores the laws that have been written. Think of it this way: if I hire a couple of guys with automatic weapons, break into a garden in San Francisco, and take, at gunpoint, fifty prize roses, I have committed a crime. If I sell them to Frank's Landscaping, and it sells them suspecting I stole them, Frank's Landscaping has also committed a crime. But if a logging contractor does the same thing in Sumatra, or Peru, and gets the wood to the United States as plywood, they're typically home-free. Same thing if they sell the logs in China and the mahogany ends up in furniture in Walmart. Illegal trade is one of the huge stories being ignored by the mainstream media. Illegal logging has been the cause of about half of all deforestation in the tropics.

I encountered this challenge in 2007. My roof deck needed replacing. The big redwood planks that the previous owner had installed thirty years ago got spongy. Redwood lumber is no longer available, or sustainable. While I was thinking about replacement, I encountered Julio Cusurichi Palacios, a Peruvian forest protection activist and Goldman Environmental Prize winner. He was on a swing through the United States to educate Americans on how their consumption patterns were driving not only deforestation but violence and crime. I learned that local residents in Peru are not the beneficiaries of the logging trade, but instead its victims. In the Peruvian mahogany industry alone, an estimated 33,000 people were working under forced labor conditions to cut trees that would later sell for thousands of dollars apiece. The illegal timber trade also has proven ties to drug smuggling, money laundering, and organized crime networks.

Frustrated that more than 80 percent of the mahogany imported from Peru into the United States was illegally sourced, Palacios and others sued six U.S. government agencies, six individuals (including the U.S. secretary of the interior), and three U.S. timber companies that were importing mahogany from hundreds of locations where Peruvian investigations had revealed that all of the logs taken were illegally harvested. The courts ruled, however, that they lacked jurisdiction to prevent the importation of contraband mahogany, since the importing companies had fraudulently obtained Peruvian customs certificates.

A few years earlier, Liberian warlord Charles Taylor, who had been found guilty of "some of the worst war crimes in history," supported his tyrannical regime with revenues from illegally logged tropical hardwoods. Only when Liberian environmental activists persuaded the UN to place a ban on the sale of Liberian tropical timber did Taylor fall from power. U.S. Secretary of State Colin Powell told the UN, "Liberia's logging industry is depleting its hardwood tropical forest on behalf of a corrupt elite and destroying an important source of the natural wealth the people of Liberia need for their own development. Charles Taylor has used revenue from the timber industry, which is now under UN sanctions, to buy arms and fuel violence throughout the region." Even then, the Bush administration was reluctant to directly ban the importation of Liberian timber, and it took years to cut off Taylor's illegal lifeline.

In 2007 Telapak, an Indonesian forest monitoring coalition, estimated that over 70 percent of timber exports were illegal, and more recent studies prepared at the request of the president of the country, Joko Widodo, estimated that the government had lost nearly $9 billion in revenues from stolen timber between 2003 and 2014.

Brazil, by curbing illegal forest clearing, came to the Copenhagen climate conference as one of the world's great success stories and climate leaders. But more recently, as the collapse of the commodity boom sent the Brazilian economy—and then Brazilian politics—into a tailspin, enforcement of the country's laws against the use of fire to clear the Amazon slumped. Alberto Setzer, coordinator of fire monitoring at Brazil's National Institute for Space Research, surveying the fires raging across Amazonia in 2016, warned they could reach unprecedented levels if authorities did not modify the existing practice of starting the fires to treat pastureland.

All this meant that rebuilding my roof deck was far more ethically perilous than I had anticipated; a lot of the wood in my local lumberyard was, in everyday terms, contraband. At the same time, as a consumer, I had the right to choose what kind of wood I purchased for my replacement deck—right? And, fresh from my encounter with the forest protection activists, and unusually equipped with knowledge about the illegal timber trade, I was both motivated and informed. Right.

Well, even for me it wasn't that simple.

I found some teak decking certified by the Forest Certification Council—a certification I trust, because it is independent of the industry. But when we got ready to build, my contractor discovered that the source had changed and that the product was no longer certified. Thankfully, he did find Forest Stewardship Council lumber still available and suitable for my deck—Iron Woods Ipe. (I also looked at using recycled plastic— much cheaper and probably the environmental first choice—but San Francisco code doesn't allow it to be used on roof decks for fire reasons.) In the end, it took quite a few hours to get to the bottom of what was truly ethical to buy.

Fortunately, a year later the United States took a big step forward in doing its part to curb the illegal global timber trade by passing legislation sponsored by Oregon Senator Ron Wyden and his colleague from Portland, Congressman Earl Blumenauer, to put teeth into U.S. restrictions on importing illegally logged timber. It's not accidental that Blumenauer and Wyden are both from Oregon. Not only does illegal timber hurt local environments, support violence and forced labor, and threaten the climate, but it also undercuts legally and properly grown U.S. timber. Oregon has a lot of that. In October 2015 Lumber Liquidators, the biggest hardwood flooring retailer in the country, pled guilty to violating this statute, agreed to pay more than $13 million in fines, and also agreed to five years of probation.

Law enforcement works, and it's good for the U.S. economy. Oregon timber farmers cheered.

What about the cost of enforcing these laws and regulations? If we made producers play by the rules and operate as good neighbors, would we be able to afford shoes and meals and houses and getting around? Or would we, to quote an infamous bumper sticker, be "Cold, Hungry, and in the Dark"?

The history of environmental regulation since 1970 shows that alarms like these come up regularly, whenever effective standards are proposed, and turn out to be poppycock. When raw materials are priced below their real cost, we waste them. When they are properly priced because they are produced responsibly, we find that we can make each tree or cubic foot of methane do a lot more work.

## WASTE NOT

To meet the needs of the billions of new consumers who are expected to join the global economy by midcentury, we will need radical innovation in how we use raw materials. We will have to produce our goods with drastically smaller quantities of everything—not only energy but other natural resources, metal, minerals, and wood, as well. The good news is that high-performing, energy-efficient, modern technology can drastically reduce both the costs and the environmental impact of manufacturing. Waste is the enemy, both of profitable companies and stable climates. Consider these facts:

Fifty percent of the hardwood used in the United States is used to make wood pallets for shipping, and half of those pallets are thrown away after one use, because of weak supply chain management. They could all be replaced by reusable, recycled plastic pallets.

An average incandescent bulb turns a minuscule 2.2 percent fraction of the coal burned to generate its power into light. LEDs are already four to five times more efficient.

A gallon of tar sands oil from Alberta uses only 10 percent of its energy content to power a car on the streets of Chicago. The rest is wasted in mining, shipping and refining the oil, internal heat losses in the engines, and idling.

So even if extracting wood, natural resources, and fossil fuels in ways that don't exacerbate climate risk does add a small premium to the cost of a board foot of lumber, a ton of coal, a gallon of oil, or a cubic foot of methane—these price increases can be overwhelmed by the economic edge of higher performance and efficiency.

Unfortunately, efficiency isn't automatic; it often doesn't happen, even when it pays. We may not recognize this, because in our everyday life we regularly encounter manufacturing innovations that ramp up performance and cut waste. Manufacturing in the United States and Europe has become remarkably more efficient and clean. If you have images in your head from 1930s movies of a steel mill, an auto assembly line, or a logging operation, and you drop by its modern replacement, your jaw drops. Elon Musk's Tesla facility is quieter than a bus station—a lot quieter. Innovation and efficiency—and their less wonderful counterpart, fewer jobs—is the order of the day in manufacturing. Computers dominate the lumber mill. Steel mills are robot encampments.

But fewer and fewer of the cars, bundles of paper, dining room tables, and steel beams that sustain your life come from advanced manufacturing facilities in the United States, Japan, and Europe. Most of them come from manufacturing havens in developing countries, where standards are weak and technology wasteful.

Since China joined the World Trade Organization, there has been an unprecedented shift in the world's manufacturing capacity from the United States, Europe, and Japan to China and other emerging markets. So large has this shift been that by 2010, 23 percent of global climate pollution was emitted in one country but the goods were consumed somewhere else.

One of the ground rules of the current trade system has been that importing nations should have no control over how something is made elsewhere—the wages paid to workers, the pollution generated, or the natural resources wasted. U.S. citizens can, and do, set safety, labor, and environmental standards for products made here, but not elsewhere. (When international agreements have established that certain practices are unacceptable, national governments can act to ban or restrict their imports—but, as we saw in the discussion of the timber trade, enforcement of international sanctions against the trade in contraband is weak at best.)

There were two reasons for limiting such "process" standards. First, developing countries worried that rich countries would use them as an excuse for restricting their imports, perhaps by saying their wages were too low. Second, the argument was framed that U.S. citizens should not be setting pollution standards in Rio de Janeiro any more than Brazilians should set emission rules in California. Yet all pollution is not local. When pollution (or other environmentally destructive activities like overfishing) affects common resources, like the oceans or the atmosphere, populations outside a country's boundaries may suffer devastating damage from someone else's weak environmental standards.

This seems obvious when, say, we are talking about water pollution in the Rhine, the Danube, or the Rio Grande. Pollution from one country can, and often does, poison the water supply of another. Yet it took a decade of diplomacy to ease the threat to Canadian lakes created by acid rain crossing the boundary from U.S. power plants, because Canada couldn't use trade sanctions to speed up U.S. action. Likewise, air pollution from China reaches East Asia—and even the western United States.

China is now establishing much more stringent domestic standards and is on a pathway to resolving much of its contribution to this problem. But other countries have not caught up. So the health of hundreds of millions of people is threatened by the failure of their neighbors to modernize their manufacturing sectors—imported goods come with imported pollution. The same thing is true of climate pollutants like

CO2, except that those pollutants threaten everyone regardless of where they are emitted. From 1995 to 2005, Europe reduced its domestic climate emissions by 6 percent, but offshoring manufacturing meant that its total climate footprint actually got 18 percent worse.

The existence of such pollution havens has slowed progress toward cleaner, higher efficiency manufacturing in all countries. Current producers are reluctant to see their domestic industries flee, because of tougher pollution standards, to lower-cost manufacturing havens. The net result is slower global climate progress.

Fortunately, this problem has a solution, which, while requiring a tweaking of the current rules for global trade, is gathering increasing support from a variety of voices. Imagine that Europe, China, the U.S., and other nations formed a Climate Protection Club that agreed to levy a $40 per ton tax on carbon to encourage efficient, low-carbon manufacturing. The WTO would permit the Climate Protection Club members, citing the Paris Agreement commitment to a low-carbon future, to levy a fee on imported steel (or automobiles containing steel) from countries without carbon taxes equivalent to that $40 per ton. This would eliminate the economic advantage of shifting the production of steel from Europe to, say, South Korea. The Korean government, if its steel is going to be taxed anyway, would almost certainly choose to institute and collect that fee itself, rather than letting Europe receive the revenues. And this would prompt reforms in the plants themselves.

Manufacturing voices (both business and labor) in Europe and the United States have advocated such a border adjustment mechanism for years. Indeed, the cap-and-trade legislation passed in 2009 by the House of Representatives envisaged such a mechanism, and when President Trump said that he was going to withdraw from the Paris Agreement, French politician Nicolas Sarkozy threatened exactly such an import fee on U.S. goods in retaliation.

Since 2009 more and more mainstream economic voices have recognized that the benefits of a global carbon tax require an incentive for all major countries to join in, and that ensuring a level field for trade in energy-intensive goods is a particularly attractive way to achieve this.

It's important to note that border adjustment mechanisms do not challenge the idea that increased trade is good for prosperity, or that trade should be conducted in a rule-based fashion. They simply accommodate the reality that because carbon abatement can have local costs, and most of the benefits are global, there is what economists call a "free rider" barrier to aligning energy performance and efficiency standards in a global economy. This is a classic form of market failure. Current trade rules simply didn't take it into account, because current trade rules evolved before there was awareness that some of the world's worst air pollution problems were global.

## RIP, HFCS

One major greenhouse gas that was long used in manufacturing is now, thankfully, on its way out. HFCs (one of the Halocarbon gases) will be gone by 2030, perhaps sooner, and, with it, a potential climate disruption of 0.5 degrees Celsius—a truly big deal. How that came to be holds lessons for how we handle other greenhouse gas pollutants.

The 1987 Montreal Protocol was signed to phase out the CFCs that were driving a hole in the ozone layer. HFCs, a related chemical, turned out to share the chemical attributes of CFCs for refrigeration, but had an insignificant impact on the ozone layer, so they rapidly took over the cooling market. Their use has been growing at 7 percent a year. When the ozone layer stabilized post-Montreal, focus shifted to climate, and scientists realized that HFCs were a Frankenstein chemical that could eventually be responsible for as much as a third of all climate change, upping temperatures by almost 0.5 degrees Celsius by 2100.

Business sensed an opportunity. DuPont, Honeywell, and Dow began patenting refrigerants that were both ozone-layer-friendly and climate-friendly. Other businesses turned to older refrigerants. India's Godrej marketed a line of propane-based refrigerators. European auto companies explored using CO2 itself as a refrigerant for air-conditioning.

Then, a year after the Paris Agreement was signed, the Montreal Protocol was amended in Kigali, Rwanda, to phase out HFCs. To acknowledge the financial impact this would have on developing countries, three schedules were put in place: one for industrial nations; a slower-track one for China and some other middle-income nations; and a third, even slower, one for India, South Asia, and the Middle East. The final schedule is not as quick as many climate advocates wanted, but opportunities remain to speed it up. The countries on the slowest track will end up investing a lot of new capital in HFC technology, which they will then need to abandon in the 2030s. With more incentives, they would likely shift to the more rapid schedule set up for China.

If the global community acts generously, India, the biggest reluctant player, is likely to agree to a faster phase-out. Here, again, the core problem is finance. If creative investment mechanisms can reduce the cost of shifting to climate-friendly refrigerants, countries like India would be willing to make that shift more rapidly.

None of this has been accomplished by cutting back on our need for refrigerators and air conditioners, nor by making them more expensive. It was the combination of private-sector innovation and public-sector regulation that is doing the job. Innovation without regulation wouldn't work, because if Indian manufacturers could continue to use HFCs indefinitely, they would behave like free riders rather than shifting to climate-friendly alternatives. But regulation would be unacceptable if U.S. and Chinese firms had not developed substitutes, and if Godrej had not taken the risk of marketing propane-based refrigerators in the face of uncertain consumer reaction.

That same combination of innovation and regulation is the one-two punch that we must use to knock out the other greenhouse gases, too.

# RESOURCE INNOVATION—THE COMING REVOLUTION

The complete replacement of HFCs without sacrificing our ability to keep cool is an example of how radical innovation is the key to making a world with billions of new consumers and higher living standards compatible with a stable climate and sustainable environment. Here's where business leadership is making the biggest difference. All over the world, manufacturing companies, whether they are German, American, Brazilian, Chinese, Finnish, Korean, Japanese, or Australian, are racing toward a third industrial revolution. The first was based on coal and steam, the second on electricity and petroleum, and the third is based on digital- and knowledge-based manufacturing, not just computerized data processing.

HFCs were the most dangerous industrial chemical for the climate, so it's vital that we are phasing them out. But ordinary cement, because of the volume we use, is another huge threat, accounting for 5–7 percent of global $CO_2$ emissions. (After water, concrete is the material used by humans in the greatest quantity: three tons a day per person.) Cement is a particular challenge because half the $CO_2$ released in making cement comes from the chemical process of treating limestone; it cannot be eliminated simply by cleaning up the energy used in heating the kiln.

If we cannot find transformational alternatives to making cement, it may prove one of the uses for which carbon capture and sequestration is most vital. But are there other ways to do the job?

Active efforts to make concrete by combining $CO_2$ from power plants with calcium and other minerals found in seawater continue, but so far these experiments are promising niches for research rather than a big-picture substitute. There is also the intriguing possibility of redeploying the recently recovered secrets of Roman cement making, which used far less limestone and heated it to a much lower temperature, while achieving higher durability by mixing in aluminum-rich volcanic ash. This could perhaps yield a 60 percent reduction in the carbon dioxide footprint of cement production.

Finally, think about how we use concrete. We throw up concrete buildings intended for use for forty to fifty years or less—with reinforcing techniques that make it effectively impossible to reuse materials when the building is replaced. Almost no effort has gone into creating construction technologies that will enable the concrete (or wood or steel, for that matter) to be reused. Investing more in better forest management and restoration, and using wood in place of cement and steel in buildings, could also make a huge difference in the climate impact of construction.

In a mid-twenty-first-century world with eight billion inhabitants, most of them now participants in the global consumption economy, it is these kinds of innovative practices that we need to encourage and accelerate across the board.

Wind turbines, for example, have been getting more efficient in part because they are getting taller, reaching heights where the winds are more constant. But taller means bigger steel towers, and there is a lot of carbon in that steel. What about flying lightweight turbine blades, effectively on kites, where the wind can keep them aloft? A number of companies are pursuing this idea, among them Altaeros in Alaska and Makani in California. It opens up the possibility of hugely "dematerialized" wind farms operating at higher altitudes with highly reliable and strong winds.

When we talk about "energy-intensive manufacturing," a lot of it involves metals—steel, aluminum, copper. And traditional manufacturing technology was "subtractive": You started out with a big chunk of steel or aluminum and then cut it down into the shape you needed, generating a lot of waste. Pouring it into a mold and casting was less wasteful, but only certain shapes lent themselves to that, and the finished product was invariably much heavier and material-intensive than an ideal structural form.

Additive manufacturing, or 3-D printing, enables computer-generated forms to be built by adding ultra-thin layers of metal or plastic in any shape or form desired. It's in widespread use only for making prototypes, but GE has begun producing nozzles for jet engines using the technique.

It enables GE to use new metals and alloys, and new shapes and forms, that previous technologies precluded. The nozzle is lighter, which saves huge amounts of waste materials.

Another approach to reducing resource waste is bio-mimicry: imitating nature. Nature does a lot of things better—consider how little material goes into a spider web, and how much area it can sweep of flies! Designers and scientists are increasingly looking at how various organisms have solved problems similar to those that face human communities, and finding ways to emulate nature's unmatched design performance. (Current techniques for producing cement from $CO_2$ were inspired by coral reefs.)

Qualcomm's Mirasol display screen, for example, mimics the light-disruptive strategy of the blue morpho butterfly to generate bright colors more efficiently than pigments. Researchers at Shanghai Jiao Tong University are developing a super-black, light-absorbing carbon film to improve the efficiency of solar panels—based on birdwing butterflies.

Additive manufacturing and bio-mimicry work well together. If you wanted to emulate the bug-capturing qualities of a spider web, and chemists had created the right sticky plastic (not so hard, I suspect), imagine trying to duplicate the web by cutting out the surrounding materials! But given the right photograph, a 3-D printer can emulate the design of any spider web, building it strand by strand, as the spider does.

So even if we think human engineering may not provide enough new ideas to meet the challenge of pushing our civilization through today's commodity and resource chokeholds and onward to climatic and environmental sustainability, we don't have to come up with all the ideas ourselves. The natural world is an enormous, largely unread, library of solutions. We need to spend more time and energy learning from it.

12

# HOW WE INVEST

*Climate change increasingly poses one of the biggest
long-term threats to investments.*
—CHRISTIANA FIGUERES, CHIEF UN NEGOTIATOR OF THE
PARIS AGREEMENT

In his free time, Mark Carney chairs the Financial Stability Board, an international network of central banks and finance ministries who work to protect the global financial system from major shocks. By day he's the governor of the Bank of England, the UK's equivalent of the U.S. Federal Reserve. Carney had previously served as governor of the Bank of Canada. Not many people get to run two different countries' central banks, but not many people have his experience and expertise, either: Carney received economic degrees from both Harvard and Oxford and worked for thirteen years at Goldman Sachs.

In 2015, Carney delivered a speech at Lloyd's of London, an insurance firm, about the future of financial markets. Instead of talking about marginal changes in interest rates or the state of the banking system, he spoke—in a room where some of the world's first insurance policies were written—about a new financial threat: climate change.

*While there is always room for scientific disagreement about climate change (as there is with any scientific issue), I have found that insurers are amongst the most determined advocates for tackling it sooner* *rather than later. And little wonder. While others have been debating the theory, you have been dealing with the reality: Since the 1980s the number of registered weather-related loss events has tripled; and inflation-adjusted insurance losses from these events have increased from an annual average of around $10 billion in the 1980s to around $50 billion over the past decade.*

*The challenges currently posed by climate change pale in significance compared with what might come. The farsighted amongst you are anticipating broader global impacts on property, migration and political stability, as well as food and water security. So why isn't more being done to address it? A classic problem in environmental economics is the tragedy of the commons. The solution to it lies in property rights and supply management. Climate change is the Tragedy of the Horizon.*

In the tragedy of the commons, people use common resources in ways that benefit them individually while imposing harm on society. Think of a fishing company that depletes an area before moving to the next, leaving nothing for local residents who rely on fish for food and income. In the tragedy of the horizon, people use resources in ways that benefit them without accounting for costs that are likely to materialize down the road. There may be no better example than climate change. Preventing it will require more than goodwill and government regulations. It will require us to employ a force that makes the world go round, but that many environmentalists have traditionally seen as an enemy: profit motive.

Some investors have tried to use the market to combat climate change by creating socially responsible investment funds. That industry has grown over the years, but it remains a niche in the global financial system, and probably always will be. Enlightened beneficence is not going to solve the problem of climate change. Only when self-interested acts are also climate-friendly acts will success be possible. In other words:

reducing carbon must offer profit opportunities for us to win the battle against climate change. This is not only possible; it's already starting to happen.

## ACCOUNTING FOR CLIMATE

The same types of self-interested incentives that have led mayors to develop green infrastructure and businesses to improve energy efficiency are also leading investors to price-in climate change risks. Climate change is still often treated as a political and environmental issue by the media, but the best reason to be optimistic about countering it doesn't come from people who work in government or advocacy organizations. It comes from people who spend their days trying to make money.

Carney understood this, and he also saw the biggest obstacle standing in the way of investors and climate-friendly investments: the opacity of the market. Which companies are most vulnerable to climate change? How big a financial impact might climate change have on a business and over what time frame? Which are best prepared? Which are taking action, and which are doing nothing? It is currently difficult or impossible for investors to answer those questions. That's a market failure. Transparency makes markets more efficient by allowing investors to value companies more accurately. The less transparency there is, the more companies are overvalued and undervalued, and the harder it is to allocate resources to maximize returns.

My company is a testament to that idea. When I started it in 1981, real-time pricing of stocks and bonds basically did not exist. Firms taking buy/sell orders could build in big spreads (commissions), because customers had no good way to compare prices. That benefited firms at the expense of their customers, but the lack of transparency also harmed firms by forcing them to make investment decisions with limited and out-of-date information. By computerizing financial data and making it available in real time, Bloomberg changed all that. Today, spreads on trades are tiny compared to what they used to be, and firms make investment decisions informed by real-time data.

Bloomberg is constantly striving to provide customers with more and better data and analysis. For instance, to help the financial industry understand the investment implications of the energy revolution, we offer Bloomberg New Energy Finance, which analyzes the industry and identifies trends within it. But now investors are seeking climate-related data beyond energy.

In the mid-2000s, while I was mayor, the company decided to begin collecting data on various sustainability measures. But data was difficult to obtain, and what data did exist was often in forms that were not comparable. Without disclosure standards that allow investors to make apples-to-apples comparisons, sustainability information can be virtually useless.

When I left city hall in 2014, I began chairing the Sustainability Accounting Standards Board, or SASB, which develops sustainability disclosure standards across different industries in the United States. The board is creating an accounting framework that helps companies determine which sustainability metrics should be reported and how. It also helps investors understand how to use that information. Our goal is to standardize disclosure practices among leading securities issuers in ways that are helpful to investors, accountants, law firms, and regulators.

Around the same time, I teamed up with Tom Steyer, the hedge fund founder-turned environmental activist, and Hank Paulson, former treasury secretary, to create the Risky Business Project. The idea is to quantify the economic risks of climate change across the United States, industry by industry and region by region. The more that business leaders and political leaders see climate change as an economic issue, not just an environmental one, the more progress we'll make—and the better off our economy will be. As Tom has said, "One side argues morality and polar bears, and the other side argues jobs. You're never going to win with polar bears."

Risky Business and SASB are focused on the United States. But the marketplace is global. So what about the rest of the world?

Three months after Carney's speech at Lloyd's, he announced the creation of a task force—led by industry leaders—to develop international standards for voluntary disclosure about climate risks. The creation of the Task Force on Climate-related Financial Disclosures was an official acknowledgment by the Financial Stability Board that environmental variables are no longer the exclusive concern of firms with a social agenda. (The board was set up after the 2008 financial crash by the world's central banks to monitor and make recommendations about the stability of the global financial system.) Instead, they are of growing concern to general investors and corporate leaders—and important to the financial system's overall health. It was the right message, and when Mark asked me to chair the group, I couldn't say no.

We assembled a great team, led by the former chair of the Securities and Exchange Commission, Mary Schapiro, who has done a phenomenal job. We brought in diverse members from across finance, insurance, industry, and government. One of the firms represented on our task force is BlackRock, the biggest asset manager in the world, with about $5 trillion in assets under management. Its CEO, Larry Fink, made headlines in 2016 by sending a letter to the CEOs of major companies outlining why he believes that the short-term thinking dominating many companies today is counterproductive. While the bulk of his letter focused on adopting better governance practices and placing less emphasis on quarterly earnings, he also urged companies to pay more attention to climate change:

*Generating sustainable returns over time requires a sharper focus . . .*
*on environmental and social factors facing companies today. These*
*issues offer both risk and opportunities, but for too long companies*

*considered them not core to their business. . . . Over the long-term, environmental, social, and governance issues—ranging from climate change to diversity to board effectiveness—have real and quantifiable financial impacts.*

While large asset managers like BlackRock have leverage to force changes in the ways firms operate through their role as shareholders, their real power comes through the signal that they send to the market by announcing that they view sustainability as a component of corporate value.

More companies are recognizing that sustainability issues affect their bottom line. Unilever—which sells everything from soap to ice cream, and also sits on our task force—has focused on how climate change could affect demand for its products. For example, as many consumers around the world face water shortages, they need shampoo that requires less water. The company responded and, by giving customers more water-efficient shampoo, has gained a competitive edge. Businesses that adapt to a changing climate, like cities, will be best positioned for success.

Demand for data on sustainability is also growing because investors are increasingly sensitive to the environmental impact companies have. Money managers who are investing pension fund money and college endowment money—big pools of capital—are demanding to know more about companies' environmental policies. They ask the management of the companies: What is your carbon footprint? What are you doing to reduce it? When big investors talk to management and indicate that this is an area of concern, you better believe that CEOs pay attention. They don't want to lose their jobs! As it is, CEO longevity is only about four to five years now. It's a tough job, with a lot of pressure and a lot of turnover. And so the investors who press them on environmental concerns really do have some clout, and that's why you see companies competing with each other to be known as leaders on climate change.

Bringing transparency to climate risks can direct capital away from carbon-intensive investments toward cleaner ones, but market transparency is only one piece of the puzzle. Other obstacles are preventing investments in a low-carbon future as well—particularly in infrastructure.

## INVESTMENT OBSTACLES

The Global Commission on the Economy and Climate, headed by former Mexican President Felipe Calderón, estimates that the world will need $90 trillion in new infrastructure by 2030. It's hard to predict the future, especially when it comes to forecasting something like infrastructure spending. But any reasonable estimate puts future infrastructure costs for the planet in the trillions of dollars, especially as middle- and low-income countries grow in wealth and population.

Building the amount of infrastructure recommended in the report would be equivalent to essentially doubling the infrastructure we have today—in the next thirty-five years. And that would require nearly doubling spending. But lack of capital is not the primary problem. Instead, it's the lack of attractive projects to invest in. Investors are seeking yield for their capital and not finding it. There are six main reasons for this.

*First, infrastructure projects require long investments that typically have solid but moderate yields.* In today's market, the steady, boring returns of infrastructure, be it for municipal bonds or clean energy, are often unattractive when compared to start-ups that hold out the promise, however unlikely, of sky-high returns.

*Second, there is a prevailing misperception that renewable energy is subsidy-driven and therefore subject to the whims of politicians.* Given the history of unpredictable regulator intervention in some places, some investors avoid the wind and solar industries altogether.

*Third, some investors shy away from anything new.* This stems from bad experiences, but also from how investment offices are structured. For example, an investment office might have a real estate portfolio but not a clean energy one. Or it might be waiting for energy investments to mature as an asset class. Investors often look for multiyear, or even decades-long, track records, making it far harder to shift into an industry that lacks that history.

*Fourth, not enough global investment flows from north to south.* This often stems from concerns about the integrity of capital markets in low- and middle-income countries. In India, Prime Minister Modi told me that the country has many entrepreneurs and companies eager to build reliable renewable power projects, but not the financing to do it. In this way, India is typical of the developing world, where capital markets are not deep and liquid enough, and interest rates are too high, to provide adequate domestic financing for clean energy.

How, for example, do we enable Japanese investors to provide capital for energy projects in India that have far higher growth potential than those in Japan if Indian stock markets are not sufficiently transparent and reliable for outsiders? Addressing such issues requires government leadership.

*Fifth, governments are still tilting in favor of fossil fuels.* The International Energy Agency estimates that, globally, governments provided $493 billion in subsidies for fossil fuels in 2014. In 2009, the G20 committed itself to phasing out some fossil fuel subsidies, but very little progress has been made.

Meanwhile, governments provide renewable energy companies with only about $120 billion in subsidies. So for every dollar that governments use to encourage the development and use of renewable energy, they spend over four dollars to encourage the development and use of fossil fuels. Even the world's major insurance companies, distressed by the failure of industrial nations to take more aggressive action on climate change, have called on G20 governments to get rid of fossil fuel subsidies.

You get what you pay for. And right now, we are paying for a hotter planet. If we are serious about slowing climate change, then we have to start putting our money where our mouth is. At the Copenhagen Climate Change Conference in 2009, OECD countries committed to delivering $100 billion per year to developing countries to help them transition to cleaner energy and adapt to the changing climate. That commitment has not yet been honored. It must be, and ending fossil fuel subsidies would allow them to do it, with billions to spare.

*And sixth, sustainable investments are often capital-intensive because they rely on technology rather than natural resources.* This makes sustainable infrastructure particularly expensive where borrowing costs are high, which is common in emerging markets where infrastructure investment is most needed. Farmers also need access to low-cost borrowing if they are going to make needed changes in our agricultural system. For instance, concentrated dairy- and beef-feeding operations are a major source of methane from manure piles and lagoons. It's simple and lucrative to convert this manure into methane to power farm machinery and even generate electricity, but it is also capital-intensive, and farmers need cheap loans to do it.

There are profits to be made from the transition to a low-carbon future, and the faster these six obstacles can be cleared out of investors' way, the sooner those profits—and the environmental benefits attached to them—can be realized.

## DOLLARS AND SENSE

Cities face their own challenges in financing infrastructure. They have a few main routes for raising revenue. Let's consider each.

**Taxes and Fees.** In 2015, residents of Seattle approved nearly $1 billion in property tax increases to pay for mass transit and safer streetscapes. Congestion pricing programs like those in London, Stockholm, and Singapore are creating dedicated revenue streams for transit improvements by charging drivers a fee to enter the busiest parts of the city. These are effective ways for cities to raise money, but as we saw with New York's congestion pricing effort, politics can often stand in the way—so cities must consider other avenues, too.

**Value Capture.** In the first decades of the twentieth century, the expansion of the New York subway system into undeveloped rural areas laid the foundation for the city's growth and enabled its transformation into the economic capital of the world. What was true then remains true in the twenty-first century: If you build it, they will come—and so will new businesses, new revenue, and rising property values. Value-capture financing essentially borrows against that rising value, allowing cities to reap some of the future benefits of their investments in infrastructure before they happen. It's a tool we used in New York to fund the city's first expansion of the subway system in more than half a century and set in motion one of the largest commercial developments in our history, in a section of Manhattan's West Side known as Hudson Yards.

Dominated by a twenty-eight-acre working rail yard, Hudson Yards was the last major undeveloped area of Manhattan. If a platform could be built atop the rail yard, there'd be room for a whole new neighborhood—right in the middle of the city. But the yard was about half a mile from the nearest subway station, a major drawback for attracting new residents and businesses. The underfunded Metropolitan Transportation Authority, a state-run agency that operates the city's subway, couldn't even keep up with repairs; building a subway extension to a sparsely populated neighborhood was not exactly at the top of the agency's to-do list. So our administration stepped in.

In 2004, we issued $2 billion worth of bonds to fund construction of an extension of the number 7 subway line and a new station at the yards. The bonds would be paid back with the property-tax revenue that would result from the development of new office and residential towers. We built it and, sure enough, they came: residents, businesses, and—

thanks to the parks we incorporated into the neighborhood—visitors from all over the city and the world.

This was hardly the first time the city used value capture to raise revenue. In fact, Central Park was financed in part through fees levied on the property owners around it. The city fathers recognized that the landowners would benefit from the park through higher property values. So if public spending was going to provide extra benefits to a particular set of landowners, it was reasonable to ask them to chip in.

If this idea had been used to build the Second Avenue subway line, which was started in 1929, it would have been completed long ago. But it hasn't been used—and as a result, the project is still far from finished. As the former head of the Metropolitan Transportation Authority, Jay Walder, noted during construction, the Second Avenue subway will reshape the Upper East Side of Manhattan: "But we don't have any mechanism to capture that back. And so we're sitting here today, as we're making progress on the first phase of the Second Avenue subway, wondering about how we fund the second phase and how we fund the third phase and how we keep this going." (The first phase finally opened in 2017.)

Many cities in Europe and East Asia make regular use of value-capture financing. Take Hong Kong. Its subway operator, MTR, doesn't only build trains, it also builds property above its train stations—apartment buildings, malls, and more. As property values rise, thanks in part to proximity to mass transit, MTR is able to capture profit and put it back into the train system. This system helps make Hong Kong's metro one of the most reliable in the world—and MTR has been winning bids to operate systems in other cities, including in Australia.

**Borrowing.** Cities have long issued bonds to raise money. Recently, this idea has expanded to focus specifically on projects that help fight climate change, through a financial instrument called green bonds. Green bonds are like any other bond, but they are used exclusively for investment in sustainability measures, from mass transit and energy efficiency to clean power and climate resilience. Green municipal bonds are usually tied to specific projects, held to specific standards, and tax-exempt, making them attractive for investors.

Institutional investors like them because they diversify their assets but also because they demonstrate a commitment to helping the planet, which is increasingly important to clients. Some money management firms even have mandates to invest in a certain proportion of assets that help address climate change. Green bonds were initially issued by international bodies like the World Bank. Now they are becoming increasingly common as a way for cities, states, and other public entities, such as transit and power authorities, to raise money. Even private companies can issue green bonds. Apple issued $1.5 billion in green bonds at the beginning of 2016 to finance clean energy and efficiency projects.

In 2014, Johannesburg became one of the first cities in the world to issue a green bond. The city raised $143 million for sustainability investments, including hybrid fuel buses. The bond was oversubscribed by 150 percent. Transport for London, which runs the metro and public buses, issued a green bond to fund cycling initiatives, expanded capacity at busy tube stations, energy-efficiency measures, and more. In 2013, $11 billion in green bonds were sold; in 2016, more than $80 billion were sold. This still represents a tiny sliver of the roughly $100 trillion global bond market. But the sliver is growing.

## CREDIT WHERE IT'S DUE

One of the obstacles cities face in raising money to support transit projects is that they often lack the authority to act without approval from regional or national governments. But sometimes the obstacle is simpler: a lack of credit.

In order to issue bonds that are attractive to investors, cities need to have good credit. However, many cities don't have a good credit rating, or any rating at all. This is particularly true in developing countries, where investment in low-carbon transportation is urgently needed and where governments are the most cash-strapped.

The World Bank estimates that only 4 percent of the five hundred largest cities in developing countries have an internationally recognized credit rating, and only 20 percent have domestic credit ratings. This effectively cuts them off from widely available capital to finance infrastructure. It's a fixable problem and a great opportunity, if public and private entities work together.

A few years ago, Lima, Peru, worked with the World Bank and other agencies to secure an enhanced credit rating. Armed with that rating, the city was able to borrow $130 million to upgrade its bus rapid transit system—money that would have been too expensive to borrow without that credit rating. It's an incredibly cost-effective fix. The World Bank spent only $750,000 providing the technical assistance to help Peru borrow $130 million.

National governments have strong incentives to help cities make these investments. But cities must do their part, too. "There is no city too poor not to be able to afford a climate change agenda," says Shpend Ahmeti, the mayor of Pristina, Kosovo. And increasingly, that agenda must include strategies for accessing capital.

Credit works for cities the same way it works for consumers. If cities can show a history of being able to borrow and pay back, lenders will be more compelled to finance projects in cities.

This starts with something every city is responsible for: good governance. For instance, Kolkata, India, has been taking steps to improve how it collects revenues and manages its budget. That has helped raise the city's credit rating, which in turn has attracted more external investment, including $400 million in support for its Environmental Improvement Investment Program from the Asian Development Bank.

In Africa, the rapidly growing Ugandan capital, Kampala, crafted a strategic plan to bring more transparency and accountability to city government. Among many other measures, the city overhauled how it keeps track of property records and collects taxes. These bookkeeping details aren't sexy, but they can make all the difference in a city's ability to meet people's needs. In one year, Kampala was able to increase the revenues it collected by 86 percent. This is money the city is now able to invest in vital services. And with a stronger bottom line, and a clear commitment to better governance, the city was able to achieve an investment-grade international credit rating.

These and other examples help to drive home a critical point: Cities don't need a miracle to beat climate change. They don't need to choose between economic growth and saving the planet. These are not technological challenges. They are challenges of policy, governance, and leadership.

This was exactly the argument a group of eighty-five mayors made on the world stage at the 2016 UN Habitat III conference in Quito, Ecuador. Led by Mexico City's Miguel Angel Mancera, Barcelona's Ada Colau, and Madrid's Manuela Carmena, the mayors called for access to international climate funds and greater control of their own finances. "Cities must have direct access to funds, direct access to financial institutions," says Mayor Colau.

There are a number of other steps we can take to speed up private finance in public projects. Cities' proposals for climate investment are often written by sustainability experts, not financial professionals, which means that many projects wind up being touted for their environmental benefits, not their financial returns. As a result, lenders often don't have the information they need to decide whether to invest. Helping cities measure and explain return on investment in sustainability projects will bring more of them to life. Technical assistance—something that the Global Covenant of Mayors helps to facilitate—can often be just as important as technological breakthroughs.

I firmly believe that markets are the most effective way to organize resources in a world of scarcity. I'm an ardent supporter of capitalist institutions, and I've spent much of my career working to make them more efficient. Our society can't function without business, which means we can't solve the climate puzzle without business involvement.

Mitigating climate change and moving to a low-carbon economy will be an investment-intensive effort that presents opportunities both for savings and for economic growth. The capitalists who seek to stay stuck in a fossil fuel past forget that progress is its own economic stimulus. The old railroad barons thought they were in the train business. They weren't. They were in the transportation business, and soon the automobile displaced rail as the dominant mode of transport. Failing to recognize disruptive innovation is a mistake that corporate leaders make time and again. But while some fossil fuel companies cling to their market share, more investors and CEOs recognize that the earth is shifting beneath their feet.

The single most important development in the fight against climate change hasn't been the Paris Agreement, or the U.S. shale gas boom, or even the advancement of solar and battery technology. All have been critically important. But the most important has been that mayors, CEOs, and investors increasingly look at climate change not as a political issue but as a financial and economic one—and they recognize that there are gains to be made, and losses to be averted, by factoring climate change into the way they manage their cities, businesses, and funds.

# PART VII

# ADAPTING TO CHANGE

13

# A RESILIENT WORLD

*For generations, barrier islands, marshes, and cypress
trees as far as the eye could see protected us from hurricanes. . . . For decades the coast has been under attack from
every angle: cut by canals, starved of nutrients, and battered
by storms. . . . This attack must stop and be reversed.*
—MITCH LANDRIEU, MAYOR OF NEW ORLEANS, LOUISIANA

Throughout this book, Mike and I have emphasized that the climate crisis is not a single problem with a single cause or solution—it needs to be
viewed, analyzed, and resolved in discrete components. Of course, it's
also true that there is one atmosphere and one climate system; so it's
important to know if the individual solutions we are implementing, when
added up, are likely to restore a stable climate close to, if not precisely
like, the Holocene epoch.

It's also vital to examine, since the climate is already changing, what
steps communities need to take to get ready for the impact of new weather
patterns. That's the focus of this chapter and the next.

Gradual climate change, after all, has been part of history since
long before greenhouse pollution began exploding about two centuries
ago. But that change was slow—and temperature and precipitation
changes took place within certain limits. What is threatening about
climate change caused by people is that it is happening much faster
than natural cycles and lacks the kind of boundaries that characterized
the last 12,000 years.

Scientists have examined their models to project what kind of sea level rise, changes in heat and precipitation, and other climate risks might result from various increases in concentrations of greenhouse gases in the atmosphere. Initially they set a redline for warming at a 2 degrees Celsius increase. Above that, it would be difficult for many regions to adapt to the climate disruption they would experience, and the chance of major catastrophes, like the sudden melting of the Greenland and Antarctic ice sheets, would increase. The current consensus, however, is that serious trouble could begin at the point that temperatures rise only 1.5 degrees Celsius (2.7 degrees Fahrenheit).

Scientists have also estimated that adding another 1,000 gigatons (billion tons) of $CO_2$ to the atmosphere (or adding its equivalent in $CO_2$, methane, black carbon, and other greenhouse gases) will make it likely that we will go over 1.5 degrees over the long term. They call that "the carbon budget." We are using up about 50 of that 1,000-gigaton budget every year, and, based on what countries pledged in Paris, we might still be emitting that much as late as 2030. If so, we would have only 250 gigatons left for the rest of the century. That means that our 2030 target actually needs to be no larger than 40 gigatons, on our way to zero net greenhouse emissions by 2050. At our present rate, it looks as if we are headed to dangerous levels of greenhouse gas concentrations by 2050.

Now, models are models—and the real world is different. For example, for the last three years the world economy has grown significantly, but emissions have stayed flat—which wasn't supposed to happen for decades. The commitments that nations made in Paris won't kick in until 2020, but many are already implementing their pledges and moving ahead of schedule. The Marrakech Climate Change Conference, around the time of the 2016 U.S. election, saw the rest of the world setting strong new goals, with Germany submitting a plan to cut its climate footprint by 95 percent, and twenty-nine new regional and local governments (many in China) committing to similarly deep emission cuts.

So cumulative progress in the first year after the Paris Agreement is hopeful. But we have to expect setbacks as well on a global journey over thirty-five years. We're also uncertain how fast we might unleash natural climate disrupters—black carbon from forest fires or methane from permafrost. So we can't be sure we will respect the 1.5-degree or even the 2-degree redline.

## CAN WE HEAL THE CLIMATE?

What happens if we do bust the 1,000-gigaton carbon budget? We can still bring concentrations of those gases back down to a level that sustains a stable climate. We won't get back quite the climate of the last 500 years, but we can shoot for one that future generations can adjust to. We have some time—remember, climate steers slowly and sluggishly, like a ship or plane, not all at once. If we go over our emissions budget briefly but promptly start reducing concentrations, we can eventually heal the climate, even though repair will take decades to be felt.

Here as elsewhere, the key to unlocking the climate puzzle is to realize that if we handle other pieces of our business better—in this case, stewardship of vital food, timber, and water sources like wetlands, mangroves, forests, prairies, and peat bogs—we will simultaneously make major strides toward a safer climate and a cooler world. Investment in ecosystems is vital if we want them to thrive and then do what they are superbly equipped to do—suck carbon out of the atmosphere and turn it into soil and vegetation.

When scientists refer to this conversion of $CO_2$ from a gas into organic matter as negative emissions, it sounds preposterous. But it's actually not. After all, it took the intervention of an industrial civilization, wastefully extracting enormous quantities of fossil fuels, breeding billions of cattle, and using large volumes of disruptive chemicals like HFCs, to disrupt the climate in the first place. Prior to this disruption, atmospheric concentrations of CO2 periodically went down, as oceans, forests, and soils sucked more CO2 out of the air than volcanoes and fires put back in. Global CO2 levels in 1750 were lower than they had been in 1500. Only with the Industrial Revolution did they begin their steady climb. Cooling, as well as warming, is in the Earth's tool kit.

Here's how it works. Natural processes scrub black carbon and methane out of the atmosphere fast—in twenty years, both they and their warming impact on the climate are gone. HFCs and nitrous oxide hang around for a long time—but if we implement the global agreement phasing out HFCs and also phase out excessive fertilizer use, they won't take us close to the 1.5-degree redline.

That leaves CO2. It can last for a thousand years, but only if it remains in the atmosphere, and most of it doesn't. Right now, only 1 percent of stored carbon is in the atmosphere; 30 percent is in soil and vegetation; 13 percent is stored in reserves of coal, oil, and natural gas; and a whopping 56 percent is in the ocean. So once we phase out HFCs and minimize methane, black carbon, and nitrous oxide emissions, our task is to store more and more of the world's carbon not in the sky (where it disrupts climate) and not in oceans (which it causes to become acidified and damaged) but in soils and vegetation, where that CO2 is the key to beneficial results, including higher crop yields and more effective water storage. More carbon in the sky is a threat. Carbon in soils and forests is an asset.

We have phenomenal machines to do this job—they are called plants, and they exist to take CO2 out of the atmosphere and turn it into organic matter in the soil. (Certain kinds of rocks also sequester CO2 in very large quantities, but our ability to influence the pace of such mineralization is much more limited than our impact on the biosphere and plant life.) What does it take to trip the biosphere—the globe's collection of animals and plants—into a massive carbon-storing mechanism?

We simply have to protect valuable natural ecosystems like peat bogs and mangroves, start planting and growing more trees than we cut, adopt farming practices that treat soil as a primary asset rather than stripping it of nutrients and carbon, and allow grasslands to be grazed in ways that enhance, rather than destroy, their productivity. Healthier agriculture and forestry, in all their forms, are the keys to reducing atmospheric concentrations of CO2.

There are other strategies. A fair amount of scientific research is pursuing engineering strategies to "capture" CO2 even once it is released in the atmosphere. Exxon-Mobil is a partner in a major fuel cell project, which combines CO2 from flue gases with natural gas to create electricity. Whether these pilots can, in a near future, be scaled at an affordable level is uncertain, but it seems only prudent to support the needed research to explore them.

One thing is certain: The use of plants to reduce climate risk is ready and able to go—today. Let me take you through the natural mechanisms that could be our salvation.

## THE KEY STEPS TO CLIMATE RESTORATION

Let's start with a familiar carbon sink, mangroves. They protect the coastline of many tropical nations from severe storms: half of the coastline in Mexico and India vulnerable to storm surge; more than a quarter in Indonesia, Myanmar, Mozambique, and the Philippines. Mangroves also provide a variety of fish and other seafood with juvenile nurseries. And they suck an enormous amount of carbon out of the atmosphere.

But mangrove swamps have a weakness. You can't charge for living near one. Regardless of whoever pays for safeguarding or planting the groves, everyone on the coastline benefits. No one investor or industry has an incentive to safeguard them, because everyone benefits equally, whether they invest or not. So in spite of these benefits, we have been losing 1 percent of mangroves each year in the Asia-Pacific region. Fortunately, the destruction rate has begun to fall in most areas, as governments recognize their value for coastal protection and fisheries enhancement. In regions where we are no longer losing mangroves, we can start restoring them. Restoration is so inexpensive in most tropical countries that the return on investment in better fishing alone is a staggering 10 to 1.

Additionally, there are huge carbon benefits. Each hectare of mangrove forest in West Papua, Indonesia, can store an incredible 2,500 tons of $CO_2$ per year. If we simply restored half of the mangroves we have lost since 1980, we would store 6 billion tons of $CO_2$, equal to total U.S. emissions each year. (We would also protect almost 2 million hectares of heavily populated tropical coastline from typhoons.)

Nature has an ingenious variety of ways to convert $CO_2$ into valuable resources; mangroves are only one example. Another, peat bogs, amount to 3 percent of the world's landmass. They store more carbon than the world's more extensive forests and grasslands. But they are easily disrupted by fires, particularly in Indonesia, Russia, and Canada, where they can burn uncontrollably and cause dangerous air pollution. Protecting them is essential.

Forests make the same conversions, and many cities and nations are putting them to work. As part of its Paris pledge, India committed to increase forest carbon storage by increasing the country's forest cover from 21 percent to 33 percent. The program was launched with a signature Indian big bang, a single-day effort to plant fifty million trees in the country's most populous state, Uttar Pradesh. If the trees reach maturity, the carbon stored in them—and, more important, in the associated vegetation and forest soils that surround them— will increase for a long time. If India reaches its goal, it will eventually store an additional 14 gigatons of carbon in its forests—as much as the country currently emits in six years. And, as the government understands, the

immediate benefits of forest restoration far exceed the costs, just looking at local dividends such as watershed protection, clean air, and forest livelihoods.

India is not alone in its ambitious forest restoration goals. In one of
the boldest developing-country climate initiatives, Kenya has announced
plans to reforest 9 percent of its landmass, an area as large as Costa Rica.
This will more than double Kenya's forest area and will help restore
watersheds that used to feed the country's hydroelectric projects but no
longer store adequate water.

One of Africa's poorest and driest countries, Niger, is making major
efforts at ecosystem restoration. In spite of its abysmally weak government, bottom-up grass roots action has succeeded in restoring 5 million
hectares of desert for trees and agriculture. The farmers used innovative,
low-tech practices such as "half-moon" pits to store water and paid
careful attention to the root systems of drought-stressed trees. The average investment: less than $20 per hectare.

Finally, agricultural soils can suck up enormous amounts of carbon.
Preliminary experiments suggest that modifying agricultural practices
to emphasize soil carbon sequestration offers enormous potential for
negative greenhouse emissions. New approaches to increase soil carbon
content are continually being tested, as discussed in the chapter on agriculture. One 2016 study estimated that annual increases in soil carbon
storage could equal 80 percent of today's fossil fuel consumption, if we
optimize such practices as reducing overfertilization, relying more heavily
on no-till farming, and using charcoal-based composts.

What do all these ecosystem opportunities to reduce atmospheric
$CO_2$ concentrations into safer ranges add up to? The Stockholm Environment Institute calculated that, taking into account a wide range of
tools to increase soil and vegetation carbon storage—protecting peat,
ending deforestation, restoring degraded forests, mangroves, and pastures,
and adopting climate-friendly agricultural policies—it should be feasi-
ble to store 370–480 gigatons of additional $CO_2$ in soils and vegetation.

Since the remaining carbon budget is only 1,000 gigatons, this is a very important buffer zone if emission reduction lags, or if the climate turns out to be more sensitive than we suspect to greenhouse gases. The combination of fast action to reduce emissions and global commitments to ongoing investments in forests and ecosystems offers the best pathway to minimizing climate risk.

Natural ecosystem enhancements combined with emissions reductions are the key to keeping long-term $CO_2$ concentrations at tolerably safe levels. But they won't do so immediately, or completely. Greenhouse emissions have been large enough that the climate has already changed, oceans are already rising, and communities are going to have to prepare for more extreme weather. And here, too, it turns out, the key player is not new technologies invented in Silicon Valley—although they will have roles—but, once again, free services we get from nature that we have forgotten to value sufficiently.

## THE RIVER'S GOING TO DO WHAT THE RIVER'S GOING TO DO

One of the most striking examples of how natural defenses can help us cope with climate change comes from the American city that has come to symbolize the perils of extreme weather: New Orleans.

In October 2009, four years after Katrina, I went to New Orleans for the Great Mississippi River symposium. The chair, Bartholomew I, the "green" ecumenical patriarch of the Orthodox Church, wore black robes and a silver cross. The speaker, Lt. General Robert L. Van Antwerp, head of the Army Corps of Engineers, wore military fatigues. The main outlet of the Mississippi River was flowing placidly past our conference room windows, a mild 600,000 cubic feet per second.

The general's fatigues reminded us that he and the corps are fighting a war with the river. The general's remarks gently prepared us for the fact that he knows he will lose that war. The outlet under our windows, he told us, can handle only 1 million of the 3 million cubic feet of water that come down the river *each second* in the spring flood. At peak, two-thirds of the river finds another outlet to the Gulf—the true main mouth of the Mississippi is now northwest of New Orleans, in the Atchafalaya Basin.

That's why the general expected to lose: because the Mississippi River is seeking a shorter path to the sea. New Orleans is no longer where the Mississippi wants to go. Effectively, the Atchafalaya is where the river yearns to escape with every spring flood. Eventually it will break loose. The Port of New Orleans will silt up.

If we don't want to lose more of Louisiana than the deepwater Port of New Orleans, we will need to respond to the reality that the Gulf of Mexico is rising, while south Louisiana is sinking. The Mississippi River, America's biggest experiment in conquering nature, is now going to test whether we can learn from experience and work with nature rather than against it—even as changing climate makes nature a far more mercurial partner.

The oldest parts of New Orleans survived Katrina best, because the first French settlers built on higher ground. Indeed, for most of human history, people built, whenever they could, on higher ground. They didn't need an environmental impact statement to advise them. It was common sense. They knew how big nature was and how important natural defenses were, so they used them. This common sense began to go astray in the mid-nineteenth century. It needs to come back.

Fifty percent of the wetland-creating silt that once flowed down the Mississippi River system is today clogging up federal dams like Ft. Peck on the Missouri, which flows into the Mississippi north of St. Louis. This eliminates more than half of the value of those dams as hydroelectric projects. It also makes it virtually impossible for the Big Muddy's sediments to reach Louisiana's wetlands in the quantities that could compensate for the gradual rise of sea level in south Louisiana. So the wetlands that once protected New Orleans from hurricanes surging up from the Caribbean are washing away. South Louisiana is losing land to the sea faster than any other place on Earth.

So north of St. Louis we need to manage the Missouri dams to route the sediment around them and back into the river. If we don't, southern Louisiana will continue losing its best defense against rising sea levels and the floods they bring.

North and south of St. Louis, the entire man-made levee system has made the river too narrow for the amount of water it must handle. During heavy rains and storms, water that would have naturally overflowed into floodplains instead stays in the river—until it crests the levees. During spring floods of 2007, two years after Hurricane Katrina caused so much damage to New Orleans, the levees north of St. Louis broke. While that was bad for soybean fields in the river's natural floodplains, it was fortunate for St. Louis. Had the levees held, the water would have continued rushing toward St. Louis and crested there—and another American city would have faced devastating destruction.

The Army Corps of Engineers, working with local and state governments, needs to dramatically and safely expand the river's natural floodplain. Otherwise, if the northern levees protecting soybean fields hold, St. Louis floods. Farther south, if cotton doesn't get flooded, New Orleans will.

Simply rebuilding or raising levees protecting New Orleans won't solve the problem because it will create a similar situation during storms: By preventing water from naturally overflowing in southern Louisiana, it will just force storm surge coming up from the Gulf of Mexico north and into the state of Mississippi and cause flooding there. The river, and the sea, are going somewhere. The answer is to restore wetlands and floodplains so the water has somewhere safe to go.

Doing that requires reversing established approaches. The Gulf Coast could potentially nest behind a three-tiered system of protection in which restored barrier islands would absorb storm surge to protect coastal wetlands and cypress forests; coastal wetlands and cypress forests would escort floodwaters onto croplands; while levees would protect communities rather than cotton fields. *Nothing else will work*—and this means that the corps, the shipping community, and the oil and gas industry all need to get out of the nature-destruction business and into the habitat-restoration business. Oil and gas operators, for instance, could do most of their work with hovercraft. Because they aren't required to, however, they instead keep dredging channels that destroy wetlands and funnel storm surge right into New Orleans and other coastal communities. Change will require considerable political will.

One of the potential silver linings from the oil well explosion at BP's Deepwater Horizon drilling platform in 2010 was that a major portion of the penalties was set aside to help finance wetlands restoration and make southern Louisiana more resilient. Projects are underway, but most of them are overly reliant on expensive engineering solutions—transporting silt by pipeline and grading it with construction equipment, for example, instead of letting the river and the tides create wetlands naturally. As a result, restoration is not taking place at the needed scale. These funds should be redirected to a more holistic and ecologically based restoration of the entire Missouri/Mississippi ecosystem, one that gets more silt back in the river, opens up its floodplains to handle high water, and eliminates dredging and other activities that accelerate the collapse of southern Louisiana into the Gulf.

## THE RESILIENCE DILEMMA

All over the world there are places like New Orleans that need new approaches to cope with the reality that the future climate will be different from today's, and probably less stable. We will need greater resilience at all scales—local, regional, and global—to cope with changes in the climate that are already occurring as we write this book.

The kind of careful urban planning and design that Mike discusses in the next chapter can help cities and towns withstand increasingly unpredictable and changeable weather at the small scale. But to protect large landscapes we need to rely primarily upon natural mechanisms. *Increasing resilience at the regional level begins with a search for the natural defenses that historically provided protection.* We must invest in strengthening those natural processes, like the Mississippi River's silt conveyor belt that created south Louisiana, instead of relying primarily on complex and hard-edged engineering projects, to ensure our safety and security.

The British learned this the hard way a long time ago, when, as the emerging colonial power in India, they decided that the port city of Calcutta was inconveniently distant from the sea, far up the winding and mangrove-choked Hooghly River.

Believing that man could conquer nature, these Victorians decided they would move Calcutta from its inland location behind the mangrove islands of the Sundarbans—India's equivalent to Louisiana's wetlands. A few voices protested that Port Canning, the new city built fronting the sea, would be vulnerable to a typhoon. But the British Raj, in the full flush of its hubris, went ahead anyway and copied Calcutta—road for road—on the edge of the Bay of Bengal. Five years later, the typhoon came. Port Canning vanished. And the Raj scuttled back behind the mangroves to Calcutta's historic location, where the city remains today.

Florida offers us another lesson—and opportunity—to use natural solutions to the challenges of climate change. As Mike mentioned earlier, South Florida sits on porous limestone rock, the relic of ancient coral reefs. Its only water supply is rainfall, which seeps down into the pores, crevices, and sinkholes of this karst landscape. Because rain seeps into the limestone, South Florida doesn't have real rivers, but it has a huge underground aquifer, the Biscayne, fed by the sheet of water that historically flowed south from Lake Okeechobee over the Everglades, gradually seeping into the limestone as it flowed.

That's the water that keeps South Florida alive. But during the same period that the government decided to "tame" the Mississippi River, it was reengineering South Florida to "reclaim" and "drain" the Everglades for all kinds of uses, ranging from sugarcane to subdivisions. A new highway built across the peninsula in the 1920s, the Tamiami Trail, created a solid dam preventing water from flowing south, effectively cutting off a large part of South Florida's freshwater supply.

With less freshwater seeping into the Biscayne Aquifer, seawater gradually began to take its place. Wells closer to the ocean had to be shut down. Then later, with climate change, the ocean began to rise, increasing the pressure of ocean water pushing inland into the porous limestone and leaving more of the aquifer salty. Additional sea level rises of only three to five inches, which could occur within fifteen years, would have a dramatic effect on available freshwater.

This potential catastrophe has no engineering-alone solution. The freshwater dammed up behind the Tamiami Trail can't be piped south. Piping that much water would be prohibitively expensive, and there would not be enough lakes to store the water when it arrived. If you built seawalls around Miami to prevent flooding, the increased pressure of higher sea levels would drive salt water into the Biscayne Aquifer underneath such walls and create floods inland, as Mike has discussed.

The partial solution: make the Tamiami Trail a bridge, not a dam, so that water can resume its flow underneath it, creating freshwater pressure inland to counter seawater pressure from the coast. The first step toward such a solution has been taken: A mile of the trail has been turned into a bridge. This is a shadow of what is needed. The National Park Service originally proposed an eleven-mile skyway, but Congress appropriated only a pittance. Predictably, it is only when more and more of the aquifer is lost to saltwater that action on the scale needed will occur. Turning things back over to nature seems to lack the political attraction of mastering her.

California faces a similar dilemma. Climate models suggest that in a warmer world, the state may actually get more total precipitation; but far more of it will fall as rain rather than snow, and annual fluctuations will be more severe. Since the state currently stores about one-third of its water supply for free in the form of ice and snow in the Sierra Nevada mountain range, the loss of that storage capacity means that California may have more and more years of extreme water shortage.

Storing more water in dams is simply not an option; the good dam sites were taken long ago. The next round of potential reservoirs that engineers have identified would cost $9 billion and yield only 400,000 acre-feet in an average year—less than 1 percent of the state's annual usage. However, underground California has enormous aquifer capacity, at least five times as much as its surface storage capacity. Getting maximum water storage underground requires managing the land on the surface with that goal in mind. To date, however, California's leaders in agriculture and forestry, the two biggest land uses in the state, don't view water as a crop, and they are not rewarded for managing their lands to increase the yield of water.

In the case of forests, this is ironic. Most forestland in the Sierra Nevada range is managed by the U.S. Forest Service, created with the mission of sustaining water supplies as its primary focus. Unfortunately, after World War II another mission took watershed protection's place: timber supply. The Forest Service developed the Smokey the Bear "every fire down by 10:00 a.m." management strategy. But these forests had evolved with fire as part of their natural cycle. Stopping all fires left them "overstocked," choked with small trees, brush, and downed limbs. Such overstocked forests pull more water out of the soil, reducing groundwater recharge. In drought periods, and as temperatures increased with climate change, the trees also became stressed by lack of water and insect attacks.

As a result of these changes, high-intensity fires now occur in California's forests at an unprecedented rate. The combination of overstocking and intense fires is drastically reducing the ability of forestlands to recharge aquifers. The forests need a complete management overhaul, one designed to maximize water yield and adapt to warmer and less predictable climate. The Forest Service is ready to make the shift. But ideological gridlock in Congress has prevented the needed funding to begin undoing the results of decades of mismanagement.

Agriculture as a source of water storage is similarly boxed in by the past. Indeed, California water law actively discourages farmers from even very cheap measures to capture and store more rainfall underneath their farms. In most places they don't have any legal right to the water they store, and their neighbors can pump it out for use or sale. There have been some gains—the state made significant improvements in its groundwater laws in 2014—but they still fall far short in allowing markets to properly operate in this area. If farmers could profit from the water they captured, they would be motivated to view water as an asset to foster.

## FEEDING THE WORLD

However severe the impact of climate change on southern Louisiana, South Florida, and California, Americans are not in danger of starving as a result.

That is not true in much of the world. As we have seen, future crop productivity is at risk from climate change. So how do we get agriculture in developing countries—in particular, agriculture that supports small subsistence farmers, or small holders—ready for climate change?

Here again, as in other sectors, providing these farmers with knowledge, innovation, and modest sums of capital is often the key. Kheyti, a start-up in southern India, sells 2,500-square-foot greenhouses for a $280, 10 percent down payment. The greenhouses control temperature, humidity, and pests; enable farmers to grow crops with 80–90 percent less water; and increase the value of the crops grown on a small plot tenfold.

In Southeast Asia, a major drought recently afflicted hundreds of thousands of households, and climate projections indicate that the region is likely to be one of the most heavily affected by weather patterns. The combination of climate change and the increased upstream construction of dams on the Mekong River by China has called the stability of the Mekong Delta into question. The likely solutions begin with deployment of traditional approaches to agricultural productivity, the reliance on networks of check dams and water catchments to store rainfall. But new varieties of rice better able to withstand variable weather patterns and salt water intrusion are also going to be essential—an area where global investment has fallen in the past three decades.

Africa is ground zero for the challenges of getting ready to live with a new and unstable climate. Here, too, El Niño hit farmers hard in 2015, particularly in East Africa. But the long-term trends are truly worrisome. When nations are ranked by how hard climate change will hit them (with New Zealand the luckiest, at Number 1; the United States at No. 11), African countries draw the short straw. South Africa, the continent's least vulnerable, is still almost halfway down at No. 84; Nigeria, Kenya, and Uganda are at No. 147, No. 154, and No. 160 out of 180 nations ranked.

Africa's climate is already marginal for agriculture, and it shows. East Africa has higher rates of hunger than any other region—one-third of the population is malnourished. Crop yields are 10 percent of those in the West, and Africa is the only region where food production per capita is already falling. The dryland regions of Africa's Sahel, the dry belt south of the Sahara, will face significant declines in available water, even as higher temperatures mean that crops will require more water to reach maturity. (Most of the water a plant uses is to air-condition itself against heat.) African crop yields could fall by 20 percent for rice, wheat, and corn; drought could slash the growing season in dry regions by 40 percent.

Africa's recurrent droughts and crop failures are rarely surprises. Weather reporting and satellite data normally signal—weeks or even months in advance—that crops are at risk of failing. If food relief arrives immediately, most of the damage can be avoided. But if it is even six weeks late, families are forced to kill livestock, pull girls out of school, and sell next year's seed corn. International relief efforts are almost never timely.

To remedy this shortcoming, African nations created, by treaty, the

African Risk Capacity program, a risk insurance pool that includes thirty-two nations and focuses on drought. ARC's objective "is to capitalize on the natural diversification of weather risk across Africa, allowing countries to manage their risk as a group in a financially efficient manner in order to respond to probable but uncertain risks." The European Union provided seed capital, and each participating African nation then chose the level of risk protection it desired and provided the appropriate premium. The scheme is designed to be commercially sustainable, using standard commercial techniques like insurance and investment of premium.

It's an important lesson in how innovative financial mechanisms can help the world prepare for unstable climate. But Africa cannot self-insure against the full force of the climate challenges it faces. Based on the pledges submitted in Paris, the continent faces a $488 billion shortfall over the next fifteen years to pay for the costs of needed resiliency.

## A GLOBAL EMBRAPA?

In the last twenty years we have seen a dismaying drop in global investment in public sector agricultural research—the kind that gave us the first Green Revolution. We need a new agricultural revolution, one designed to enable farmers, particularly small holders in the tropics, to thrive in a less stable, and in many cases more challenging, climate. We need, in fact, a global Embrapa.

Embrapa is the Brazilian Agricultural Research Corporation, which made Brazil the first tropical agricultural superpower. Like the institutions that created the two previous scientific agricultural revolutions—the 1940s hybrid corn revolution in the United States and the 1970s Green Revolution around the world—Embrapa is open source (i.e., publicly accessible) and publicly funded. It tested and disseminated a broad system of agricultural interventions suited for tropical conditions.

In the ten years after its launch, agricultural production rose by 365 percent, without genetically engineered private patents and without destroying rain forest for new cropland. (Rain forest continued to be destroyed, but for other reasons. Deforestation contributed little to the agricultural miracle.)

What did Embrapa do? First, it focused on improving soils, discovering that Brazil's *cerrado* (savanna) soils needed lots of limestone to counter their acidity. Then, using conventional cross-breeding, Embrapa created an enormously prolific variety of tropical grass, which greatly expanded Brazil's ability to grow grass-fed beef, again without destroying rain forest. Finally, again using conventional breeding, Embrapa took the classic temperate climate crop—soybeans—and made it suitable for the tropics. It pioneered no-till agriculture and created an integrated farming system using crops, livestock, and trees.

Public sector research like Embrapa's freely disseminates commonly shared agricultural systems—seeds, fertilizers, livestock, techniques. These are then deployed by markets, but created by governments. If we're going to feed a world of seven, eight, even nine billion people in a rapidly changing climate, we are going to do it with approaches like Embrapa's, not with privatized, patent-protected, proprietary approaches. (The U.S. approach to genetic engineering is an example of the latter.) They are simply too narrow for the challenge. Small holders are not lucrative customers.

But developing new technology is not the whole solution to agricultural resilience; small holders need to be able to access it. Here innovation is being pioneered by a variety of institutions. One new entrant is the Climate-Smart Lending Platform being sponsored by the Global Innovation Lab (which Bloomberg Philanthropies supports with grants). The Lending Platform will work with farmers and financial institutions to develop a variety of standardized loan products for farmers who agree to adopt new "climate-smart" agricultural practices. Modeling suggests that this combination of climate-smart practices, better measuring, and access to affordable, de-risked capital can enable farmers to increase their profits two- to fourfold.

Today our enormously enhanced ability to measure and understand natural dynamics, combined with a newly humbled awareness that, in fact, we are not able to tame even a single hurricane adequately to fully protect New York or New Orleans, gives hope for a new approach to climate. What is most exciting is that the approaches that pay off in the near term—in a higher quality of life in Manhattan, more abundant fish catches in the Philippines, cooler Aprils in New Delhi, better water quality in Palm Beach, lower insurance premiums in New Orleans—also make each of these communities more stable in the face of a disrupted twenty-first-century climate, and help assure their security.

14

# NEW NORMALS

*The measure of intelligence is the ability to change.*
—ALBERT EINSTEIN

As dawn illuminated New York on October 30, 2012, the trail of destruction left overnight by Hurricane Sandy came into full view for the
first time. I'd spent much of the previous twenty-four hours at the Office
of Emergency Management in Brooklyn. Throughout the night, we had
been getting reports on the storm's power and impact and coordinating
the city's response. On the Rockaway Peninsula, in a neighborhood called
Breezy Point, a fire destroyed more than eighty houses. In areas of south
Brooklyn and Staten Island, entire neighborhoods flooded under surging
water that reached as high as fourteen feet, leaving some people who refused to evacuate their homes stranded on the second or third floors. Homes
were knocked clear off their foundations. After flooding caused a trans-
former to explode at a power plant on Fourteenth Street near the East River,
the power went out in much of Manhattan. The Empire State Building,
and nearly every home and office building south of Thirty-ninth
Street, was dark. Subway tunnels flooded, paralyzing much of the city's
mass transit network. We had never seen a storm like this in New York.

That morning, I went to Breezy Point to survey the destruction, meet
with families, and assure them that we would do everything possible to

help them. Farther up the Rockaway Peninsula, I walked down Beach Boulevard, where each year I'd marched in the community's St. Patrick's Day parade. It was hard to believe it was the same place. Personal possessions that had washed out of homes littered the streets. Sand, mud, and smashed cars covered the sidewalks. Anguish and disbelief hung in the air. I'd see the same kind of wreckage, physical and emotional, in neighborhoods across the city.

Sandy claimed the lives of forty-three people in our city, and in the days after the storm, we took every possible step to prevent that toll from rising. While first responders continued search-and-rescue operations, building inspectors worked around the clock to make sure homes and businesses were safe to enter, and to let people know not to enter those that weren't.

The Department of Sanitation had to remove hundreds of thousands of tons of debris to open roads. With public schools shut down, children and parents had their lives thrown out of balance. Many cell phone towers had lost power, so service was out in much of the city. In the hardest-hit neighborhoods, people were cut off from fresh food and water—some of them elderly people stranded on high floors of apartment buildings. We set up temporary centers to distribute emergency supplies, including to the more than six thousand people staying in emergency evacuation shelters, many of whose homes had been damaged or destroyed. Thousands more had to be evacuated from hospitals, nursing homes, and health care facilities, some of which had their power sources in flooded basements, including their backup generators. Thanks to heroic work by many, not a single patient's life was lost.

In the midst of this work, some people debated whether Sandy was caused by climate change, but that missed the point. Ocean temperatures and sea levels are indisputably on the rise, and both changes have the potential to magnify the force and frequency of storms—along with the risks we face from them. The first job of city leaders is to protect people, and Sandy made clear that New York was vulnerable in ways we couldn't delay addressing. So even as we recovered from Sandy, we began working to prepare for the future and all the possibilities that it might bring.

A month after Sandy, we brought together business, political, and civic leaders to announce that we'd create a blueprint for building a stronger, more resilient city. Thankfully, PlaNYC had given us a big head start.

## REMEMBERING CANUTE

The Dutch have an old saying: God made the world, but the Dutch made Holland. There's a lot of truth to it. The name Netherlands means "lower lands," and they are just that: More than 20 percent of the coastal nation lies below sea level and another 50 percent is less than one meter above sea level. Its viability as a flourishing, advanced nation would not be possible without an extraordinary network of defenses built over the centuries. Following a 1916 storm that flooded parts of North Holland, the Dutch began constructing a system of sea protections that is one of the great marvels of civil engineering. After devastating flooding in 1953, the Dutch expanded the system to better protect southwestern areas of the country. The Dutch system of coastal protections was built gradually, piece by piece, and that work continues today in response to rising sea levels.

The Dutch example holds a valuable lesson: We can't ignore the risks we face from Mother Nature, but we also don't have to run from them. With smart planning, and political will to invest in the future, we can manage risks, adapt to them, and thrive alongside them.

We began this work in New York City early in my administration. In launching PlaNYC in 2007, I asked every agency to begin taking steps to make their operations more resilient to extreme weather. We engaged with the public, utilities, telecommunications and transportation companies, and regulators about how to fortify our city for the changing climate. The NYC Panel on Climate Change assembled the most comprehensive local projections for climate change in any city in the world. And we adopted new zoning rules that required people to take steps to protect their homes and businesses from floods. We also required climate risk assessment for major development projects in vulnerable areas. For that reason, almost all of the city's new waterfront development got through Sandy pretty well. This included new buildings along the East River in some of the city's fastest-growing areas, like Williamsburg and Long Island City. New parks we'd built, from Brooklyn Bridge Park to Governors Island, took rising seas and coastal storms into account in their construction, and they made it through the storm with minimal damage.

In fact, after taking stock of the storm's destruction, one figure jumped out: Of the eight hundred buildings that were severely damaged or destroyed during Sandy, 95 percent were more than fifty years old, and built before modern building code standards were put in place. In other words, virtually every structure that was built according to modern standards survived the storm in reasonably good shape.

From the moment we announced PlaNYC in 2007, we were constantly defending the idea that New York City should be investing to protect itself from climate change. In 2009, while New York—and the rest of the world—was fighting to rebound from the global recession, we held an event in the Rockaways to announce the latest findings of the New York City Panel on Climate Change. We aimed to use the data to make strategic investments in resilience to protect people's homes and livelihoods. For this, I was ridiculed in the press for following the advice of "hot-head scientists" who wanted to "push climate ahead of jobs." But being a mayor requires a willingness to look further ahead and take criticism from those who rarely see past the tip of their nose.

Three years later, Hurricane Sandy put climate change back on the media's radar. In the weeks and months after the storm, the press was hungry for solutions, but the answer that media outlets obsessed over—a giant seawall!—couldn't have been more wrong. Building a massive seawall was no more practical than Donald Trump's plan to build a huge wall across the Mexican border. In politics, and too often in journalism, simplicity sells. But the truth was: a massive seawall would take forever to construct and would be prohibitively expensive. By the time it was completed, rising sea levels might render it obsolete. Complex problems rarely have simple solutions.

Consider this: if the storm had hit New York a few hours earlier or later, because of the tides, it would have flooded entirely different parts of the city—like parts of the Bronx and northern Queens near the Long Island Sound and the East River—and rendered any huge wall across the harbor practically useless. At the same time, while Sandy's storm surges were responsible for most of its impacts, the storm brought relatively little rain or high winds. If it had brought more of either, the city would have faced a whole new set of issues.

There's always a tendency to fight the last war. After World War I, France built a fortified line of defense along its eastern border, the Maginot Line, where Germany had invaded. It was, in essence, a wall designed to prevent the advance of enemy troops—but in World War II, Germany simply went around it. In responding to Sandy, we were not going to build the climate change equivalent of the Maginot Line.

Our job was not to prevent another Sandy but to make the city stronger and more resilient to the wide variety of evolving challenges we faced. That included not only coastal storms but also extreme heat, heavy rains, and severe droughts—all of which are becoming more frequent as the planet warms, and none of which can be solved by a giant wall.

I was open to any and all ideas, so long as they were achievable. I wasn't interested in enormous projects that looked great on paper and could theoretically solve every problem but that stood no chance of actually getting done. And I had no interest in trying to tame Mother Nature.

Humans have been trying to stop the tides from coming in since King Canute. As the legend goes, Canute's subjects believed his powers to be omnipotent, which frustrated him. To show them that even a king had limits, he had his throne brought to the edge of the sea and ordered the waves not to break on his kingdom's shores. Of course, the tides still rolled in. The press teased me for citing that fable after Sandy, but it still holds a lesson for us: Human power is no match for Mother Nature. The challenge we face is not walling ourselves off from the elements but learning to live with them—and adapting to change, including rising and warming seas.

That doesn't mean we have to retreat from the waterfront. In New York City, it's one of our greatest assets. Yet, for decades, New York had neglected it, and it had become polluted, degraded, and abandoned. Our administration had spent eleven years reversing that history and revitalizing the waterfront, opening it up for all New Yorkers to enjoy.

People want to live near the water. They always have, and, if you ask me, they always will. Human history has been shaped by our relationship to the water in so many ways. On every continent, national economies are anchored by port cities, and New York's harbor is the reason the city became a global economic capital. The challenge is to learn from each extreme weather experience and better prepare for the next. After Hurricane Irene in 2011, for example, we expanded the mandatory evacuation zones we'd established in 2006 to include new neighborhoods; that adaptation may have saved lives during Sandy.

It is true that, in some places, economics and science will dictate that the wisest course is away from the coast. After Sandy, New York State created a voluntary buyout program that some homeowners signed on to—including an entire community in Oakwood Beach, Staten Island. The houses were sold to the state at an agreed-upon price. They were then knocked down, and the land is being returned to nature. New Jersey has also created buyout programs in endangered beach communities. And the federal government has begun buying out communities that are in danger and can't afford to defend themselves. Interestingly, two of these threatened communities—Isle de Jean Charles, Louisiana and Shishmaref, Alaska—are on opposite ends of North America, showing how widespread the effects of climate change have become.

Of course, many people don't want to move, and others can't afford to. And an agricultural community of two hundred people is different from a modern city of ten or twenty million. More than 500 million square feet of buildings lie in FEMA's one-hundred-year floodplain for New York City, which is equivalent to the entire city of Minneapolis. In 2012, they were home to almost 400,000 people and more than 270,000 jobs. We can't pick them up and move them—and we don't have to.

## NATURAL ALLIES

Sandy made clear how important modern building standards are to protecting our cities. The storm also drove home a point many coastal communities have long known: Our best defense against nature is often nature itself.

Grassy coastal wetlands slow waves and reduce the force of their impact when they reach shore. Beaches and dunes put space and barriers between the water and shoreline development. Elsewhere in the world, as Carl has explained, saltwater mangroves provide tropical cities with an even stronger first line of defense. Mangroves can absorb up to 90 percent of normal wave energy, and their protection can mean the difference between life and death.

During Sandy, the Rockaways provided a clear picture of the effectiveness of natural defenses. Beach Ninety-fourth Street was not fronted by dunes, and it was heavily damaged. Just two miles down the shore,

Beach Fifty-sixth Street, safeguarded by dunes, was mostly fine. In 2006, the city of Long Beach, just a few miles up the Long Island coast from the Rockaways, debated a plan to build dunes along its waterfront. Proponents cited the protection the dunes would offer, while opponents worried that they would block views of the ocean and change the character of the city's historic boardwalk. The plan was eventually rejected by the Long Beach City Council.

When Sandy struck, it hit Long Beach hard. Homes and businesses were flooded, and many were lost. The famous boardwalk, touted by some preservationists as a reason for rejecting the dune plan, was so damaged that it had to be torn down. Meanwhile, in nearby communities like Point Lookout and Lido Beach, fifteen-foot protective dunes had slowed surging waters and prevented severe damage. The same lesson could be found around Long Island and New Jersey: Where dunes had been built, communities had been spared the worst.

Unfortunately, as cities have grown, our natural defenses have been dismantled. Dunes get removed as cities develop along the water's edge. Natural wetlands are lost to landfill and development. Since 1900, 50 percent of the world's wetlands have been lost. Over the course of New York City's development, 90 percent of our wetlands were destroyed.

We began reversing this process through PlaNYC. We started restoring the aquatic habitats that have provided natural protection and other benefits to New York since long before the first people arrived. We even helped engineer a comeback for a key player in New York's history: the oyster.

When the Dutch arrived in the early years of the seventeenth century, there were hundreds of acres of oyster beds in New York Harbor. By some estimates, our waters held half the world's supply. Oysters were a staple of early New Yorkers' diets, just as they had been for our Native American predecessors, and their shells were a critical source of fertilizer and construction mortar. Liberty Island (home of the Statue of Liberty) and Ellis

Island, both synonymous with the values New York stands for, were originally called Great Oyster Island and Little Oyster Island. Throughout the eighteenth and nineteenth centuries, our docks were filled with oyster barges transporting the mollusks around the United States and overseas, and oyster carts were as common in our streets as hot dog carts are today.

Well, sure, you may say—but what does that have to do with climate change? Well, oysters are natural filters that clean water by removing impurities. Oyster beds also protect coasts by slowing waves and currents. Over time, due to overharvesting and increasingly polluted waters, New York's oyster population plummeted—and studies have shown that those decreases coincided with increases in coastal erosion. As part of our work restoring the city's natural defenses, we began reintroducing oyster beds and restoring habitats to help them thrive. We've greatly reduced pollution in New York's waterways, but am I ready to enjoy a dozen East River oysters on the half shell? Maybe someday.

Of course, oysters alone can't protect a city from climate change; they are just one small piece of a larger puzzle. When you put those pieces together, though, they can make a big difference. After Sandy, we proposed a new network of natural defenses: everything from berms and walls to planted dunes and revitalized wetlands. And these features wouldn't just defend the city—they'd also make it more beautiful and vibrant. A flood wall doesn't have to be just a wall. It *shouldn't* be— especially in cities like New York, where every square foot of space counts. A barrier that prevents flooding can also be a public park, planted with grass and trees and lined with paths and playing fields. It can even be a new neighborhood.

New York's Battery Park City was built on landfill along the Hudson River starting in the late 1970s, and developers were required to take storm preparation into account. The neighborhood is raised above water level and buildings are set back from the water by a ring of parks and green spaces. During Sandy, the West Side of Lower Manhattan fared far better than the East Side, and Battery Park City was a big reason. If it hadn't been there, flooding in the Financial District and into the World Trade Center site might have been much worse.

So development can actually *decrease* the risks cities face—but only if it's done wisely. After Sandy, I proposed that future administrations consider the idea of doing for the East Side of Lower Manhattan what Battery Park City had done for the West Side: Build it out, raise it above the flood level, and develop it. We called the concept Seaport City, because of its proximity to the historic South Street Seaport. Yes, it would be expensive to build. But over time it would pay for itself, just as Battery Park City has done. Demand for housing and office space in Lower Manhattan has never been stronger. Seaport City could capitalize on that demand, bringing thousands of new residents and hundreds of businesses to the area while also protecting existing residents and businesses from future storms. Whether or not future mayors pursue it, I hope they think just as big—or bigger.

## MAPPING OUT SOLUTIONS

As I mentioned in Chapter Four, communities around America have long gauged their flood risk by FEMA's flood maps, which also outline the contours of the federal government's National Flood Insurance Program. The program, created by Congress in 1968, provides lower-cost insurance for homeowners living in areas at elevated risk of flooding. The federal government requires anyone with a federally backed mortgage who lives in a flood zone to have flood insurance—and most choose to buy it from the federal government. There are more than 5 million such policyholders in the United States. Florida has the most, with nearly 2 million. Texas, Louisiana, California, and New Jersey round out the top five.

These maps affect people's lives, and, in large part, they are woefully outdated. That means that people don't have information they need to make informed decisions about where to live. It means that many people who should have flood insurance don't, and it means that premiums aren't aligned with actual risks. Without updated maps, new developments in coastal areas aren't required to be built to high standards. Creating these maps comes with a price tag, but it pales in comparison to the costs we'll incur if we continue leaving communities unprotected.

In 2007, as we were creating PlaNYC, we formally requested that FEMA update its New York City flood maps. It delayed, because it didn't have the data, so we had to get creative. Another federal program, aimed at identifying optimal locations for solar panels, provided funding for taking precise fly-over measurements of building and street dimensions with laser imaging, or LIDAR. By participating in the solar mapping program, we were able to create an up-to-date topographical map of New York City. We required the contractors gathering the data to ensure that it match and exceed FEMA standards. (The federal government should have been doing that in the first place, so that every city creating a solar map could also get an updated flood map.) We then handed over the data to FEMA, which enabled it to update versions of New York City's flood maps.

This data turned out to have many other important uses as well. The maps helped the NYPD manage public events. They also helped us gauge the health of our wetlands. But most communities around America don't have the means to collect their own mapping data. States and Congress should help them.

Bringing FEMA maps up to date would mean that millions more homes and businesses with federally backed mortgages would be required to have flood insurance. It would also increase premiums for many people who have flood insurance, because their level of risk would rise. So as we update the maps, we must ensure that flood insurance costs don't force people from their homes.

One way to do that is by reducing premiums for those who take precautions to protect themselves. For the most part, the federal program will reduce rates only if homeowners raise their homes. If you do raise your house, your premium will fall considerably. But that doesn't work in New York. Many of our buildings can't physically be elevated. People who take other precautions—like raising boilers and electrical systems out of the floodplain—should also be able to get a reduced rate. We proposed that change after Sandy, but FEMA and Congress have yet to take action. FEMA is also required by law to explore creating a program to subsidize insurance costs for low-income homeowners. Four years after Sandy, and more than a decade after Katrina, no such program exists.

There are also many people who aren't required to have insurance but probably should be. After Sandy we proposed a program for low-cost, high-deductible policies so that more people would be covered in case of an emergency. Congress has yet to adopt that, either.

Cities can't wait for national governments to act. They must find their own ways to help people protect themselves affordably. For instance, homes and businesses in vulnerable areas could pay into programs that fortify local defenses across a given neighborhood. Because this would reduce risks, it would likely reduce insurance premiums, making the fees worthwhile. Cities also have the authority to decide what kinds of structures are allowed in particular areas, and they could consider allowing more development in exchange for private investment in coastal resilience. Under such an arrangement, cities could allow developers or businesses to build taller if they agree to help fund coastal defenses. The steps that cities and towns take to reduce the risks facing homes and businesses, like the ones we proposed after Hurricane Sandy, should also be taken into account in National Flood Insurance Program rates. Doing so would incentivize local governments to make smart investments that protect citizens' homes and pocketbooks—and encourage residents to hold their leaders accountable when they don't.

Building resilience requires cities to coordinate with other local and regional government agencies. For instance, cities often don't own the systems that provide power, although their residents depend on them. New York's utilities are private companies that provide a public service, often using public assets to do so. They must be held accountable for keeping the power on, especially during an emergency. In New York, utilities have to meet benchmarks for reliability, but those benchmarks didn't include preparation for natural disasters, and they didn't take climate change into account. After Sandy, I directed our team to assess exactly what it would take to make every essential network capable of withstanding a major hurricane, or a record-breaking heat wave or other natural disaster. What would it take to get there, and how much would it cost? You can't build a road if you don't know where you're going, and without knowing what's possible, you can't set priorities.

In our plan after Sandy, we recommended more than 250 separate achievable measures. The total cost of everything we proposed was just over $19 billion. It's a large sum. But the total damage New York City suffered from Sandy was almost exactly that: $19 billion. And that figure could be much larger the next time a major storm hits.

According to a 2005 study by the National Institute of Building Sciences, every dollar invested in resilience saves four in avoided costs. After Sandy, our projections showed that, if we did nothing, a storm of similar intensity hitting New York in 2050 could cost around $90 billion, thanks to rising sea levels.

Even if climate change were to miraculously freeze in its tracks, the investments we proposed would be worth it. They would protect communities from normal periodic flooding that occurs even during minor storms, as well as from extreme weather. They would leave our city better protected for the future but also a better place to live today, and they would create jobs and strengthen the city's economy. With some government intervention and expenditure up front, cities and nations can save significantly in the long run. But the less we do now, the bigger the role government will have to play later.

The lessons we learned can help other cities avoid disaster before it strikes. After the storm, we worked to spread those lessons through city networks like C40, which helps cities organize into communities based on the shared challenges they face. So, for instance, through the Connecting Delta Cities network, places like New Orleans, Ho Chi Minh City, and Rotterdam—all of which are particularly vulnerable to flooding—can share experiences and wisdom. We all have a lot to learn from one another.

## RAIN

Rising seas aren't the only challenge facing cities. Around the world, sudden and severe rainstorms are becoming more frequent, as rising temperatures speed evaporation, causing moisture to gather more quickly in the atmosphere.

In the summer of 2016, torrential rains unleashed severe flooding across cities in China. More than 160 people were killed, tens of thousands of houses were destroyed, and economic losses exceeded $30 billion. Historically, China's cities were better built to handle rain and flooding, but rapid urbanization has destroyed many of the natural systems that kept water out of cities and in rivers, lakes, and ponds. As the scientist Vaclav Smil pointed out in his book, *Making the Modern World*, China used more cement between 2011 and 2013 than the United States used in the entire twentieth century. Worsening flooding is one dangerous legacy of that development.

The number of Chinese cities experiencing flooding has doubled since 2008. In 2013, more than two hundred Chinese cities experienced flooding at some point. At the same time, China is grappling with the challenge of providing water for its expanding population—many of its largest cities face water shortages.

Chinese cities are now adopting an ambitious idea to help address both problems: creating "sponge zones" that absorb water. This includes building large parks that absorb stormwater and using highly permeable concrete that allows water to pass through paved surfaces to areas where it can be collected and reused. This concrete is fascinating to see in action. It looks like regular concrete, except that you can dump a cement truck full of water on it and the water instantly disappears, as if being poured down a drain. China isn't the only city experimenting with solutions like this. Los Angeles is also beginning to soft-surface some areas, so that the rainfall will recharge aquifers instead of flooding neighborhoods.

In Rotterdam, where 80 percent of the city lies below sea level, another interesting solution to the problem of severe rain is being tested: building *down*. The city's "water square" is a sunken public space that can be used for recreation and relaxation the vast majority of the time—but during heavy rains, it can collect up to 450,000 gallons of water channeled from surrounding streets and rooftops. The water then filters down into the soil or is pumped into canals. The park's basins and gutters even double during dry weather as skateboard ramps.

Rotterdam aims to be "climate-proof" by 2025. Its strategy isn't to encase the city in a bubble but to make climate change manageable through smart infrastructure. In one neighborhood, all the homes have moved electric wiring to upper floors and out of the way of flooding. The city is subsidizing the installation of green roofs and considering other financial incentives, such as tax breaks for residents who install systems to collect and recycle rainwater.

Every inch of rain in New York City equals one billion gallons of stormwater—and that can add up to big problems. In August 2007, heavy rains shut down much of New York's subway system. In response, we began fortifying the city against the impacts of major rainstorms. Part of this work included overhauling stormwater management—an even less sexy topic than retrofitting buildings. It didn't make a lot of headlines, but it improved the city in important ways.

Briefly: like many cities, New York has a combined sewer system—when it rains, water from the streets goes down into sewers and mixes with sewage wastewater. In normal weather, all of this water is treated at wastewater plants, then discharged clean into bays and rivers. But during severe rainstorms, sewage treatment plants can't always handle the combination, and untreated sewage winds up in waterways.

It is an engineering problem that comes out of an unfortunate by-product of development. As cities grow, they transform nature's permeable (and carbon-absorbing) surfaces into impermeable concrete, stone, and steel. The green infrastructure plan we adopted in New York City as part of PlaNYC, like that of cities around the world, is reversing that process in small doses by making the whole cityscape more permeable. Instead of the water going into the sewer, it's absorbed by soil, plants, and stone. We installed thousands of absorbent sidewalk extensions called bioswales that channel water from streets into soil and rock. Most passersby don't notice them. If they do, it's because they are planted with trees and flowers and pretty to look at.

We also provided grants to help private property owners install planted roofs. Not only can these roofs catch rain, but they also help reduce heat in buildings, which reduces carbon, air pollution, and heating bills. We redesigned public playgrounds so that they could capture and store rainwater as well.

These steps—and other common sense measures, such as manually checking storm drains for blockage when heavy rains are expected—helped to make New York's waterways cleaner than they'd been in decades. They also greatly reduced our costs for processing wastewater.

When we hear about making cities "greener," people are usually referring to lessening impacts of climate change by reducing emissions. But we can also do a lot to adapt to changes by greening in a different way, by helping our cities behave more like natural landscapes.

# HEAT

We can also learn from nature in taking on a risk facing cities in every part of the world: heat.

To counter the heat island effect described in Chapter Seven, cities are returning to their roots, if you will—by planting trees. Trees create shade and absorb heat. They also absorb CO2. In New York City, a public-private partnership with actress Bette Midler's New York Restoration Project allowed us to plant one million trees across the five boroughs. We planted tree number one in the South Bronx in 2007. Bette and I were joined by a junior high school chorus, some third graders who read poems, and—who else—Big Bird from *Sesame Street*.

New York isn't alone in planting trees. Cities from Melbourne to Mumbai to Madrid are doing so because it's an affordable and effective way to combat climate change, and it makes cities more beautiful and healthy places to live. It all gives new life to the proverb: The best time to plant a tree was twenty years ago. The second best time is now.

# DISEASE

One of the more startling climate change stories from 2016 had to do with infectious diseases. In July, thawing permafrost uncovered the carcasses of reindeer that had been killed more than seventy-five years ago by a bacterial disease, anthrax. The warm weather reactivated the disease, sickening dozens of people and causing at least one death. Diseased and deceased deer are not at the top of the list of risks we face, but the story illustrates the unpredictable ways warmer weather is changing the world.

As regions become warmer, infectious diseases spread to new areas, and outbreaks can become more severe. What's more, flooding and heavy rainfall can also lead to more contaminated water supplies, which could lead to more gastrointestinal diseases like cholera and dysentery. Health departments and city governments need to understand these potential changes and take steps to protect people. Many already are. For instance, São Paulo's climate change adaptation plans recognize how rising temperatures can increase the incidence of dengue and other infectious diseases.

Climate risks aren't always as cinematic as rising seas and violent storms. But that doesn't make them any less dangerous. Hurricane Sandy claimed the lives of forty-three New Yorkers. A tropical city that fails to adapt to rising temperatures could lose many more lives to infectious disease. Simple measures can be taken, such as improving drainage in areas where mosquitos breed or requiring screens in windows.

## DROUGHT

One of the most serious risks facing many cities is not the prospect of more water but less. Because droughts are likely to increase in the years ahead, water management is one of the most critical areas of adaptation that growing cities face. One-quarter of the world's population already faces water scarcity.

As Carl touched on in the previous chapter, rainwater is not only one of the most valuable resources on the planet but also one of the most wasted. Capturing more rainwater will go a long way toward addressing shortages that come from droughts and depleted water sources, be they glaciers, lakes, or underground aquifers.

For the most part, cities rely on centralized systems of water distribution and collection—it's piped in from reservoirs and piped out through sewers. This system has enabled civilization to flourish, but it's not necessarily optimal. The water we need to drink and the water we need for other activities—from putting out fires to flushing toilets—comes in together in the same pipes and usually goes out together in the same pipes, because individual buildings have no way to separate and reuse it. That's in part because of regulations that govern water systems, which protect public health.

San Francisco is working to change this paradigm. The city's Non-potable Water Program streamlined the permitting process for buildings to collect, treat, and recycle water from rain, kitchen and bathroom appliances, and other sources. It then provided grants to help buildings install such systems, which can reduce demand for outside drinking water by up to 50 percent for residential buildings and an incredible 95 percent for commercial buildings. The city then passed requirements for buildings over a certain size (or that use a certain amount of water) to have water recycling systems, the first such law in the nation.

The city's program diverts water from the sewer system and also saves building owners money. And it is saving millions of gallons of increasingly precious drinking water. San Francisco's program, under the leadership of Mayor Ed Lee, was driven by the urgency of the state's historic drought. But such systems can help every city, and San Francisco is now working to help spread them. The city is leading a nationwide effort to identify institutional and regulatory barriers that keep cities from being able to adopt decentralized water systems. The next step is to create guidelines for policies that cities can adopt that benefit consumers and also utilities.

Innovative ideas to reduce waste can be applied to energy as well as water. If you've ever lived or worked in a city, you've probably noticed, maybe with some frustration, that at the same time in the same city, some places need to be cooled and others heated. For example, an office building that's being heated to keep workers comfortable in winter also houses rooms full of servers that need constant air-conditioning—because data centers give off a lot of heat and operate better at low temperatures.

A French tech company, Stimergy, found a way to address this imbalance: by recycling the excess heat generated by its data centers. In

Lyon, Stimergy is using a data center to heat a public gym. Elsewhere it is installing data centers that double as "digital boilers" to heat apartment buildings. And it has formed a partnership with Paris to heat a public swimming pool using underground data centers. (Data centers consume 3 percent of global energy—and counting—and a large portion of their operational costs come from their cooling systems.)

It's a promising concept with a wide range of applications. And it's a theme we see over and over again: A big part of fighting climate change comes down to simply making the systems that guide our lives work better.

A month after Sandy, I laid out our vision for the city's future, one in which we faced a new reality shaped by a changing climate. I recalled how the city we know today exists only because of how New Yorkers responded when faced with adversity in the past. The Great Fire of 1835 burned much of Lower Manhattan to the ground in part because firefighters didn't have access to an adequate water supply; in response, the city and state dammed the Croton River and built the aqueduct system that still delivers some of the world's best drinking water. The creation of our modern subway system was spurred by the Great Blizzard of 1888, which paralyzed our elevated trains and brought the city to a standstill. And after one of the deadliest industrial accidents in American history, the Triangle Shirtwaist Fire of 1911, New York created health and fire safety codes that became national models.

No one who rides the New York subway or turns the tap to pour a glass of water thinks of the challenges that led to the creation of our critical infrastructure. If the world's cities respond to climate change with the same determination and vision as New York's early leaders did, it will yield technological and social marvels that will improve people's lives in profound ways—so profound, in fact, that they will begin to be taken for granted. Many of the adaptation measures being pioneered today—like streets that absorb stormwater—fix problems that cities have faced since long before climate change became a concern. Future generations may not even think of them as responses to looming threats but as common sense features of any smart, thriving city.

15

# THE WAY FORWARD

*We know that climate action only works when we get
everyone involved: our government, our businesses,
neighborhoods, and residents.*
—MARTY WALSH, MAYOR OF BOSTON, MASSACHUSETTS

We can stop global warming. Not by slowing down economies but by speeding them up. Not by depending on national governments but by empowering cities, businesses, and citizens. Not by scaring people about the future but by showing them the immediate benefits of taking action. If we accomplish this, we will be healthier and wealthier. We'll live longer and better lives. We'll have less poverty and political instability. And, while we're at it, we'll pass on to our children and theirs a brighter future.

If taking on climate change is so attractive, why haven't we done it more effectively? Throughout this book we have seen how problems often result not from the economic costs of climate protection but from the failure of markets to reflect the economic advantages of action. Fixing market failures is neither a new nor an impossible challenge. Working together, businesses, governments, and civil society are already repairing many of these market flaws. We simply need to do more of it faster. Markets, like gardens, feed and enrich us. But they also require tending. Weeds need to be pulled out and fertilizer added in. Here is how we can do it.

**Reform Subsidies.** In life, you often get what you pay for. Nations currently pay enormous subsidies to fossil fuel producers and large agricultural interests, thereby overheating the climate, slowing down energy innovation, distorting agricultural markets, depleting natural resources, increasing poverty, and imposing enormous health costs on communities. On top of all that, fossil fuel subsidies deprive governments of funds that could otherwise be used to finance a transition toward a sustainable, climate-friendly future—or to lower taxes. When governments give fossil fuel companies lower tax rates or cheaper loans or other kinds of special favors, they give those companies a competitive advantage over cleaner forms of energy. Reforming these budget-draining subsidies would provide an enormous portion of the $90 trillion necessary to finance the sustainable infrastructure we will need by 2050.

**Increase Transparency.** The more transparent markets are, the more efficiently they perform, the more productively resources are allocated, and the more participants are able to hedge against risk. Transparency requires reliable data, and data on companies' exposure to climate change has long been lacking. This is now changing, through entities like Bloomberg New Energy Finance and other analytic services, the U.S. Sustainability Accounting Standards Board, and the climate disclosure task force of the Financial Stability Board. For markets to drive our climate policy goals, transparency is essential so that capital will flow to investments that minimize exposure to climate risk. For this reason, we need all economic sectors—including fossil fuel companies, manufacturers, commodity traders, banks, insurance companies, and government regulators—to measure and disclose data on climate-related risks. With the right metrics and indicators, firms can start managing risk, and markets can price in the true costs of inaction on climate change.

**Force Monopolies to Compete.** Markets thrive with competition. They fail with monopolies. On occasion, a government-sponsored monopoly makes sense. For instance, we give utilities monopolies on running electric wires through neighborhoods. But there is no good reason for giving them monopolies on producing and selling electricity. Anyone who produces electricity—from solar panels, for instance—ought to be able to use power lines for a fee. One doesn't have to own a railroad in order to ship something via its rails. Why should one have to own a utility to sell electricity via its wires? And yet in many states and countries, laws prevent home and business owners from doing just that.

Other government-sponsored monopolies, such as those granted through long-term patents, can also impede progress. For instance, patents that slow the spread of genetically modified crops that are more drought-resilient can cost lives. We need adequate public investment in a wide variety of open-source innovation in sustainable technologies. Government can underwrite the research, and private sector leaders all over the world can implement the most promising results.

Informal monopolies also impede competition. The oil industry has an informal monopoly through its ownership of gas stations. How eager do you think oil companies are to add charging stations for EVs? Manufacturers of EVs haven't been able to sell them to a mass market in part because customers fear that they won't be able to charge up. Oil's dominance of transportation energy derives not from its low price but from its lack of competition. That's why cities have been creating charging stations.

**Invest in Natural Resources.** We all rely upon natural services that we don't pay for—sunshine for light, plants for oxygen, rainfall for water, soils for food. When we let industry dump its wastes into these common resources, we all pay the price—in disease or death. To protect the public, we regulate how common resources can be used—but too often, we don't invest in those resources. Although everyone benefits from cleaner air or more abundant fisheries or better protection against storms from mangroves and wetlands, no one has sufficient personal incentive to invest in improving them. One of the keys to solving the climate problem is to increase our public investments in them—and use incentives to encourage the private sector to do so, too. If farmers who follow regenerative agricultural practices that store more carbon in their fields don't get paid for that benefit, they won't be likely to make the shift.

**Realign Incentives.** Some profitable investments aren't made because the investor can't collect the profits. Earlier, we examined the case of landlords who don't pay the utility bills and therefore have no incentive to improve energy efficiency. Similarly, shipping companies don't replace inefficient freighters because the customer pays the fuel bills. These market failures can be solved by private actions, but they may need public enabling. "Green leases," which allow the landlord to collect utility bill savings, are used in some cities. But often government needs to set regulatory standards—requiring apartments to be upgraded for energy efficiency when a tenant moves out, say, or establishing minimum efficiency standards for new cars and ships.

Other profitable investments don't happen because the investor wants to maximize the life span of its capital stock—factories, machinery, technology, and other resources. If these kinds of assets remain efficient and productive as they age, that's good. But if investments in newer and better resources are being slowed or stopped by government subsidies, regulation, private monopolies, or other market barriers, that's bad. It's bad for the economy, because it hurts productivity—think of a factory that produces less than it could and at higher costs. It's also bad for the environment, because it prevents the deployment of sustainable infrastructure. Government can help solve this problem by eliminating those market failures, and by offering companies incentives to invest to replace aging factories and machinery.

**Improve Liquidity.** Days before 174 nations and the EU signed the Paris Agreement in New York, a Bloomberg News headline read: "Wind and Solar Are Crushing Fossil Fuels." Indeed, investment in new wind and solar electricity was doubling investment in gas and coal. But that same week, Secretary-General Ban Ki-moon warned: "Sustainable, renewable energy is growing, but not quickly enough to meet expected energy demand." So what's the problem?

As we've seen with solar panels, energy-efficient buildings, and other sustainable investments, climate-friendly infrastructure is typically cheaper to operate than the traditional models but more expensive to build, because the technology has to be paid for up front. Capital-intensive projects—solar panels or LED lights, for instance—usually involve borrowing, but interest rates in many of the countries most in need of such investment are high (often in the double digits) and capital markets are thin.

True, we have lots of capital seeking yield. But loaning money to developing countries to be repaid over twenty years creates risks unrelated to the project itself. What will happen to exchange rates? Will the government be stable enough to ensure that contracts are honored? Will there be a war or revolution? Multilateral development banks, led by the World Bank Group, are the logical institutions to reduce risk in sustainable infrastructure investments in developing markets. By doing so, they will give investors more confidence that they will be able to reap returns. The importance of tackling this obstacle cannot be overstated, and it is perhaps the biggest challenge threatening the climate.

**Crack Down on Rent Seeking.** "Rent seeking" is an economics term that refers to acquiring special benefits without paying for them. Most of the lobbying industrial complex in Washington is based on rent seeking: a corporation seeking to write a tax loophole into a bill, or an industry seeking to obtain subsidies or other advantages that no other industry enjoys. Lobbying is a $3 billion industry in Washington alone—and that's not counting the lobbying that goes on in state capitals and city halls. Coal companies are rent seeking when they ask President Trump to resume below-market coal leasing in the Powder River Basin. When new technologies emerge on the scene and threaten to disrupt industries, the old guard tries to secure through politics what it cannot do through the market, as utilities have done in trying to stop homeowners with rooftop solar panels from putting excess energy into the grid. Of course, this phenomenon is not limited to the United States. It is universal, and it is holding back progress on the transition to renewable energy and sustainable economic growth.

## FIXING POLITICAL FAILURES

Market failures have not been the only impediment to action. Political failures have also played a major role. Particularly in the United States, but also somewhat in Canada, Australia, and other nations with abundant fossil fuels, climate denial has slowed down national responses.

In the United States, some of the climate skepticism is faith-based: The chairman of the U.S. Senate Committee on Environment and Public Works, James Inhofe, Republican of Oklahoma, wrote *The Greatest Hoax: How the Global Warming Conspiracy Threatens Your Future*. He has grounded his position in religious terms: "The hoax is that there are some people who are so arrogant to think that they are so powerful, they can change climate." Popes Francis, Benedict XVI, and John Paul II— along with many other religious leaders, including evangelical leaders like Rick Warren—disagree. Throughout history, when faith and science collide, it can take decades—even centuries—for the new knowledge to be assimilated, as ongoing battles over evolution attest. The difference this time is that the battle is not just over knowledge but what that knowledge requires: action.

The good news is that the religious movement for action is growing, as more faith leaders rally around the belief that it is our responsibility to care for God's creation—"creation care," it's often called. The World Council of Churches, which represents more than a half-billion people around the world, has divested from fossil fuel companies. Pastors in Appalachia are helping to lead the fight against coal. Polls show that a majority of Christians recognize that there is a scientific consensus that human activity is warming the planet. Still, there is a significant minority that believes otherwise. It's critical that we find ways to reach them. That can mean building alliances between environmental and religious groups—or not. The Evangelical Environmental Network organizes the faithful but avoids groups like the Sierra Club—"a bunch of weirdos," its leader once called them. (It takes all kinds.)

Much more than religion, climate change skeptics and deniers in the United States often hold their position based on partisan reasons (many Republican primary voters view climate change as a Democratic issue); ideological reasons (their goal is to shrink government, and climate change requires more government action); and tactical reasons (they depend on fossil fuel companies and other special interests for campaign donations).

Nevertheless, the conservative case for action on climate change is clear and convincing.

First, as we have seen, many of the government actions that are most necessary—like opening utilities up to competition from solar panel owners and ending subsidies for fossil fuels—require nothing more than applying the free market principles that conservatives champion.

Second, the investments in infrastructure that reduce emissions also make the United States more economically competitive. A study by the McKinsey Global Institute found that infrastructure investments generate a rate of return, in economic expansion and larger tax revenues, of 20 percent. Even if the rate is one-third of that, the investments would pay for themselves. Historically, Republicans have led the way in building much of the nation's most important infrastructure, from the transcontinental railroad to the Hoover Dam to the Interstate Highway System. These and other investments in America's transportation and energy networks were essential to our development as an economic superpower. We need conservatives to renew their commitment to a limited *but vigorous* government that creates the conditions for commerce to increase and businesses to grow.

Third, we don't know with certainty all the future effects of climate change, but it would be reckless to ignore the possibility that the changes will be costly, and to not take steps to mitigate them. Rolling the dice by taking a wait-and-see approach is betting the farm—and the planet— on long odds. There's nothing conservative about that. Being conservative means being cautious about the future. It means taking steps now to mitigate events that, if they come to pass, will exact a terrible toll, both in lives and dollars lost.

Fourth, the heart of the word "conservatism" is to conserve. Today, this most commonly involves conserving public dollars and traditional cultural norms. But historically, it has also meant conserving natural resources. The conservation movement was long led by Republicans like Theodore Roosevelt, who created 150 national forests, 51 federal bird reserves, 4 national game preserves, 5 national parks, and 18 national monuments. It was a Republican Congress that passed the Antiquities Act (1906), allowing Roosevelt to dedicate those national monuments. Richard Nixon signed the Clean Air Act, along with other pieces of environmental legislation, including a bill creating the EPA. And while the father of the modern conservative movement, Senator Barry Goldwater of Arizona, occasionally clashed with EPA regulators, he remained an ardent conservationist: "While I am a great believer in the free enterprise system and all that it entails," he wrote, "I am an even stronger believer in the right of our people to live in a clean and pollution-free environment."

Fifth, and finally, if we do nothing, our children and theirs will pay for our shortsighted and selfish disregard for the future. What is conservative about burning through a child's college fund? Theodore Roosevelt summed it up nicely: "Of all the questions which can come before this nation, short of the actual preservation of its existence in a great war, there is none which compares in importance with the great central task of leaving this land even a better land for our descendants than it is for us."

Some conservatives recognize this. George Shultz, secretary of state in the Reagan administration, played an instrumental role in persuading President Reagan to overrule cabinet members opposed to an international treaty that phased out ozone-depleting CFC gases. He remains a strong supporter of action on climate change, and more conservatives are joining him. The president of a libertarian think tank, Jerry Taylor, strongly supports a carbon tax, pointing out that there is no "right-wing theory of atmospheric chemistry." As the Republican mayor of San Diego, Kevin Faulconer, says: "I don't look at climate change through a partisan lens. God knows there's enough partisanship on the national level. I look at it from a quality-of-life standpoint."

Of course, liberals have their own share of blind spots, including opposing wind farms that might "spoil" their ocean or country views. "Not in my backyard" politics affects both left and right, but nothing gives climate skeptics more ammunition than when liberals refuse to practice what they preach.

Liberals also tend to oppose an energy source that is emission-free: nuclear. The Obama administration's Clean Power Plan did not allow states to count existing nuclear plants toward their emission-reduction mandates, which makes no sense, as Obama's own former energy secretary, Steven Chu, pointed out. In 2015, there were 99 nuclear reactors operating in the United States, generating nearly 20 percent of the electricity we use. But as old plants close in the years ahead, few are being built to replace them. Only two are currently under construction; they will be the first in the last three decades.

Two of the factors driving opposition to climate change action in the United States have been just as strong, and sometimes stronger, in much of the developing world. First, there has been an expectation that low-carbon energy and other green policies are highly expensive. That was once true, but no longer. As we have outlined, the cost of renewable energy is often cheaper than fossil fuels, but the up-front capital costs can be prohibitive, which is why governments must work with the private sector on financing tools that overcome those obstacles. And second, special interests in other countries are acting in the same way as special interests in the United States, clinging to their privileges, monopolies, and market positions. Logging interests in Peru, coal interests in Australia, cattle interests in Brazil, owners of outmoded merchant ships—all try to slow progress toward a cleaner world by extracting political concessions from governments.

These special interests will hold on for as long as they can, and many will succeed in extending their profitability far longer than the market would naturally allow. They will oppose the kind of market-driven reforms we have proposed that could enable the world to finance $90 trillion of sustainable infrastructure: ending fossil fuel and agricultural subsidies, cracking down on pollution and resource theft, and breaking monopolies and opening markets. We should invest some of the resulting revenue in natural commons and public goods like research, and then use private sector investment tools to finance the rest. Political leadership is essential to deploy these ideas with the speed and scale we need.

## THE METROPOLITAN SOLUTION

Historically, nations relied on their agricultural and mineral resource base and viewed cities as unreliable partners. As a result of these traditions, national governments, with exceptions like Singapore, have distributed power based on geography, not population. Most representative democracies have voting systems weighted—some might say rigged—to give more power to rural areas than population warrants. The United States is the perfect example of this. The Senate and the Electoral College are both weighted to favor rural states with small populations. The states that are more urban pay far more in federal taxes than they receive in federal program benefits. States with more rural populations pay less and receive more. The same is true in other countries.

This tradition is challenged by the reality that twenty-first-century economies are rapidly substituting knowledge and information for raw materials and commodities. Cities, not rural areas, are the nexus of this new economy—and cities, as we have shown, also have enormous incentives to take action on climate change.

Since markets are centered in urban areas, it makes sense for urban areas to lead the way in addressing market failures. And, fortunately, cities are uniquely equipped to do it.

Goods and services that benefit everyone require a collective mechanism to finance them: government. City walls were one of the first public goods that governments were charged with providing. But close behind them were streets and port facilities, drought-proof water supplies, public marketplaces to lower the costs of trading, garrisons, courts, and commercial regulation, including standardized weights and measures. These and other public goods have been provided by and in cities since their inception.

Modern cities and towns are now responsible for providing a far broader array of public goods: education, waste removal and sewers, snow removal, storm drains, clean drinking water, police and fire protection, street lighting, parks and playgrounds, communicable disease prevention, and even, in many cases, venues for entertainment and sports. In an era of unstable climate, enormous new investments will be needed in new public goods. Coastal cities should lead in expanding flood and storm protections, promoting better financing, expanding transit services, and designing buildings to minimize the urban heat island effect.

The good news is that the special interests that stand to lose from these improvements hold far less sway over mayors than they do over national legislators. The bad news is that cities often face legal obstacles in financing and implementing these solutions. Many cities don't have the power to adopt a local sales tax. Most do not have credit ratings. And some even lack the legal authority to borrow for the infrastructure they need and their citizens want.

Where cities have the most authority, they have made the most progress—on making buildings and waste treatment plants more efficient, expanding cycling and bus services, and even on generating power. Every U.S. city that has gone outside of its monopoly utility provider to obtain power has been able to get electricity that was both cheaper and cleaner. Yet most cities globally are forced to buy electricity from a monopoly, however dirty it may be, and at whatever price and reliability it provides.

Cities also need more control over their streets. Globally, transportation patterns are shifting dramatically. Many cities aim to have more of their travel provided by mass transit, and new ride-sharing services are challenging traditional ways of getting around. Autonomous vehicles are coming, but will they result in more congestion or less? Cities need the authority to manage their transportation systems in ways that allow them to reduce pollution, improve health, and combat climate change.

With 70 percent of the world's $90 trillion infrastructure bill centered on urban areas, it's also crucial that cities are treated equitably in their access to financial tools. Right now, for example, the World Bank can only make loans to nations. Yet many of the world's cities are larger than dozens of smaller member states of the UN, which routinely get World Bank loans.

Cities are eager to provide the kind of transparency and accountability that lenders and investors appropriately expect. Members of the newly launched Global Covenant of Mayors, representing more than 7,000 cities, have agreed to report their emissions and climate progress according to a standard set of tools that are more rigorous than those currently used by many countries.

Everyone who cares about climate change—and improving public health, increasing economic growth, and raising standards of living—should urge their national governments to devolve more power to cities. It is starting to happen, but too slowly. More than any national law or policy, devolving power to cities is the single best step that nations can take to improve their ability to fight climate change and, with it, the health of their citizens and economies.

This is not to say that national governments are not important. They are. Only they can address some of the market failures that stand in the way of progress, such as agricultural and energy subsidies that privilege fossil fuels. Only they can make sure that patent laws don't get in the way of the spread of climate-friendly technologies. Collaboration among nations, like collaboration among cities, can help spread innovative policies and make it easier to adopt multinational agreements. In addition, heads of state hold the keys to multinational organizations (like the World Bank) that have a vital role to play in expanding access to financing.

But how much can cities accomplish in the absence of robust national leadership? How much can they change? The answer is: a lot. New York City is a great illustration. Forty years ago, it was in a state of disaster—nearly bankrupt and falling apart at the seams. Congress and the White House refused to come to the rescue, with the *Daily News* blaring on its front page: FORD TO CITY: DROP DEAD.

Instead, slowly but surely, and thanks to the foresight and dedication of many civic leaders inside and outside of government, New York City began a comeback that still astonishes all of us who remember the bad old days. Back then, few imagined that the city was capable of the extraordinary rebirth that is now so visible—in waterfront parks full of children; in safe streets where people are not afraid to walk at night; in revitalized neighborhoods with new shops and rebuilt homes; in air that is cleaner than it's been in fifty years; in more extensive mass transit links that have pushed subway ridership to record highs and yes, even in beach chairs in the middle of Times Square. City leadership—elected officials, business leaders, and citizens—made it all possible, by working together and in partnership with other levels of government.

## OPPORTUNITY FOR ALL

There is another major area where all levels of government must flex their muscles: helping those who lose jobs, and towns that lose tax revenue, in the transition to renewable energy and a sustainable economy. Our parents' generation had the opportunity to live middle-class lives without obtaining college degrees or developing advanced skills. That has become increasingly hard to do. Technology has disrupted nearly every industry. Factories that once required hundreds of hands are now run by a few that manage robots. Construction projects that once would have required thousands of workers now require only a fraction of that. Computers have reduced the need for clerical staff, researchers, and workers in many other fields. Jobs in growing industries tend to require college degrees or advanced skills, but only about one in three American adults holds a bachelor's degree or higher.

A strong and resilient middle class—one of America's greatest contributions to the world—is being hollowed out before our eyes, and those in Washington have watched it happen. This has hurt millions of Americans, our sense of unity and optimism, and our ability to fight climate change. We need mechanisms designed to make rapid change fairer to everyone and more palatable to those connected with dying industries. Accepting disruptive change is not easy, but governments can help ease resistance and diminish fears.

This is not just an American challenge, and it is an enormously difficult and complex one for governments to confront, but confront it they must. The alternative—an increasingly discouraged, frustrated, pessimistic, and alienated public—puts at risk the future of western democracy and the stability and prosperity it has brought. This is not a book about that challenge, per se, but it's not one we can ignore. Climate change actions such as closing coal plants can have negative repercussions for workers, which must be confronted head-on. There are ways to address the fallout and preserve the overall benefits—and here are three ideas for action in the United States.

First, let's bring back the New Deal's original vision for the Tennessee Valley Authority and apply it to the coal belt. TVA today is primarily a publicly owned utility. But in its original form it was heavily focused on restoring the environment of the Tennessee Valley, creating thousands of jobs in the process. Today, the coal belt has been left with a legacy of ruined hillsides, clogged streambeds, polluted waterways, and dangerous or abandoned mine shafts. An enormous amount of restoration is required. The same skills that were once used to mine now-exhausted, uncompetitive coal seams—and in some cases the same equipment—could be put to service in this effort. Why not put miners and others back to work rehabilitating the land that the coal companies have left so scarred?

Second, a fee of a quarter or half cent per kilowatt hour on transmitted electricity could raise up to about $20 billion a year. This revenue could be used to fund an insurance program for former coal workers at risk of losing their pensions or health care coverage. It could support communities that have experienced job losses, and lost tax revenue, from the transition to cleaner energy—as well as communities that are left with the degradation of strip mining, coal ash dumping, and other environmental disasters created by the coal industries. Think of it as an insurance policy that would be used by those citizens and communities harmed by the transition away from fossil fuels.

At the same time, the fee could help fund the transmission upgrades that are needed to increase grid reliability and integrate larger volumes of renewable power sources, which would benefit everyone—by enhancing national competitiveness, cleaning the environment, and protecting the climate. And the customer savings from the grid improvement would outstrip its total cost.

Third, and this is an area where cities and local communities are leading: We need to pull our education system out of the twentieth century and into the twenty-first, to ensure that opportunities for good-paying jobs are spread throughout the U.S., including in coal country and other regions that are affected by the transition to clean energy.

Right now, too many schools are not preparing their students for college and careers.

Too many students from poor families who could excel at competitive colleges don't apply because they don't think they can afford it—even though, in most cases, they can, thanks to financial aid.

Too many students who don't want to attend college are left with a high school curriculum that fails to prepare them for careers.

Too many vocational programs are stuck in the seventies, disconnected from today's growing industries.

And too many students who attend community colleges are poorly served by those institutions, often leaving without diplomas or marketable job skills. Community colleges offer an enormous opportunity to bridge the skills gap between the labor force and the economy, but capitalizing on it requires that we rethink their role and hold them accountable for moving students from enrollment to graduation to careers or further study.

Each of us can do more to tackle climate change in our everyday lives. Louisville Mayor Greg Fischer has urged his constituents to "do your part. Plant trees in your yard. Consider a cool roof when you replace yours at home or your business. Take public transportation when you can. Reduce idling when you're in your car, and get out and walk and cycle more." All are important. All make a difference. But none is enough.

There are many more actions that each of us can take in our lives that, collectively, can have a major impact on the future of the planet. One of the most important is also one of the simplest: doing a better job communicating with friends, family, and neighbors—not only about the nature of the problem, but also about the benefits of the solutions. Yale's Climate Change Communication program identifies six different types of reactions that Americans have to climate change: Those who are alarmed, concerned, or cautious account for 72 percent of the country, while only 28 percent are disengaged, doubtful, or dismissive. Yet it's a mistake to dismiss that latter group or condescend to them. Trying to frighten them may be worst of all, because it may lead some to view the issue through a partisan or class lens.

The best way to reach skeptics is for more people to tell climate success stories. How taking action improves our lives in the here and now. How it makes us healthier. How it extends our life span. How it saves us money. How it makes it easier for us to get around. How it helps connect those in poverty with job opportunities. How it helps us compete in the world. How it strengthens our economy. How it helps create jobs. We will never win hearts and minds simply by trying to convince people to stop eating meat or give up their cars, but we can win them by demonstrating how fighting climate change is good for them, their families, and their communities.

This is a conversation that must be led by everyday citizens and local leaders. Those in Washington will continue to pander to the extremes so long as the extremes hold power and votes. It's up to the rest of us to change the tenor and tone of the climate discussion—away from partisanship and toward problem solving. Away from fear and toward hope. Away from ice caps and toward jobs and health. And away from Congress and toward communities.

The faster we can move in that direction, the better our odds for over-coming one of the greatest challenges—and capitalizing on one of the greatest opportunities—the world has ever known.

# ACKNOWLEDGMENTS

We are grateful to everyone who is working every day to protect our planet, many of whom made this book possible. We are lucky to have a terrific environmental team at Bloomberg led by Antha Williams, Adam Freed, and Curtis Ravenel. Each of them, and many of their team members, offered their insights and expertise, including Dan Firger, Amanda Eichel, Lee Cochran, Kelly Schultz, Ailun Yang, Melissa Wright, Lee Ballin, Meridith Webster, and Jaycee Pribulsky. Others at Bloomberg Philanthropies and Associates—Kelly Henning, Janette Sadik-Khan, Seth Solomonow, and Amanda Burden—helped us strengthen the connections between climate and health, transportation, and urban planning. Allison Jaffin, Kim Molstre, Howard Wolfson, David Shipley, Jason Schechter, Tom Golden, and Doug Bernstein also lent time to the project. Nancy Cutler and Suzanne Foote helped us get to the finish line. Frank Barry, Gabe DeVries, and Scott Bade provided invaluable research and writing assistance. And it was Kevin Sheekey, with support from Patti Harris, who originally convinced us to take on this project, believing that it could advance the conversation on climate change here at home and abroad.

## ACKNOWLEDGMENTS

Leaders working in each of the three areas we focused on—cities, businesses, and communities—provided input, ideas, and data, including Mark Watts of the C40 Cities Climate Leadership Group; New York City Hall alums Rit Aggarwala, Cas Holloway, Seth Pinsky, and Marc Ricks; as well as Ben Schwegler, Carlo Ratti, Adam Wolf, Patrick Holden, Tom Heller, Brent Harris, Deborah Winshel, Carsten Jung, Graeme Pitkethly, Roger Seabrook, Lysanne Gray, Ion Yadigaroglu, Henry McLoughlin, and Jason Bade. Harish Hande and Anand Shah were enormously helpful guides to India's energy future. And we drew on superb scientific and economic studies by both public and private sector entities, particularly the energy and climate research laboratories supported by the U.S. government, and the expert team at Bloomberg New Energy Finance.

We also owe a debt of gratitude to Amy Stein, who produced our charts and graphics; Janet Byrne and Alice Truax, who assisted with editing; and the team at St. Martin's—including Sally Richardson, Jennifer Weis, Sylvan Creekmore, and Ryan Jenkins—who believed in the project from the beginning and guided us to completion.

Finally, we want to thank all the mayors, business leaders, and citizens we have met over the years who inspired us to push for greater and faster progress. May they inspire many others, too.